Advances in Industrial Control

Other titles published in this Series:

Digital Controller Implementation
and Fragility
Robert S.H. Istepanian and
James F. Whidborne (Eds.)

Optimisation of Industrial Processes
at Supervisory Level
Doris Sáez, Aldo Cipriano and
Andrzej W. Ordys

Robust Control of Diesel Ship Propulsion
Nikolaos Xiros

Hydraulic Servo-systems
Mohieddine Jelali and Andreas Kroll

Strategies for Feedback Linearisation
Freddy Garces, Victor M. Becerra,
Chandrasekhar Kambhampati and
Kevin Warwick

Robust Autonomous Guidance
Alberto Isidori, Lorenzo Marconi and
Andrea Serrani

Dynamic Modelling of Gas Turbines
Gennady G. Kulikov and Haydn A.
Thompson (Eds.)

Control of Fuel Cell Power Systems
Jay T. Pukrushpan, Anna G. Stefanopoulou
and Huei Peng

Fuzzy Logic, Identification and Predictive
Control
Jairo Espinosa, Joos Vandewalle and
Vincent Wertz

Optimal Real-time Control of Sewer
Networks
Magdalene Marinaki and Markos
Papageorgiou

Process Modelling for Control
Benoît Codrons

Computational Intelligence in Time Series
Forecasting
Ajoy K. Palit and Dobrivoje Popovic

Modelling and Control of mini-Flying
Machines
Pedro Castillo, Rogelio Lozano and
Alejandro Dzul

Rudder and Fin Ship Roll Stabilization
Tristan Perez

Hard Disk Drive Servo Systems (2nd Ed.)
Ben M. Chen, Tong H. Lee, Kemao Peng
and Venkatakrishnan Venkataramanan

Measurement, Control, and
Communication Using IEEE 1588
John Eidson

Piezoelectric Transducers for Vibration
Control and Damping
S.O. Reza Moheimani and Andrew J.
Fleming

Manufacturing Systems Control Design
Stjepan Bogdan, Frank L. Lewis, Zdenko
Kovačić and José Mireles Jr.

Windup in Control
Peter Hippe

Nonlinear H_2/H_∞ Constrained Feedback
Control
Murad Abu-Khalaf, Jie Huang and
Frank L. Lewis

Practical Grey-box Process Identification
Torsten Bohlin

Modern Supervisory and Optimal Control
Sandor Markon, Hajime Kita, Hiroshi Kise
and Thomas Bartz-Beielstein

Wind Turbine Control Systems
Fernando D. Bianchi, Hernán De Battista
and Ricardo J. Mantz

Advanced Fuzzy Logic Technologies in
Industrial Applications
Ying Bai, Hanqi Zhuang and Dali Wang
(Eds.)

Practical PID Control
Antonio Visioli

Soft Sensors for Monitoring and Control of
Industrial Processes
Luigi Fortuna, Salvatore Graziani,
Alessandro Rizzo and Maria Gabriella
Xibilia

Jie Bao and Peter L. Lee

Process Control

The Passive Systems Approach

 Springer

Jie Bao, PhD
School of Chemical Sciences
 and Engineering
The University of New South Wales
Sydney
New South Wales
Australia

Peter L. Lee, PhD
The University of South Australia
Adelaide, South Australia
Australia

British Library Cataloguing in Publication Data
Bao, Jie
   Process control : the passive systems approach. - (Advances
   in industrial control)
   1. Passivity-based control 2. Process control
   I. Title II. Lee, Peter L., 1954-
   629.8
ISBN-13: 9781846288920

Library of Congress Control Number: 2007928319

Advances in Industrial Control series ISSN 1430-9491
ISBN 978-1-84628-892-0   e-ISBN 978-1-84628-893-7   Printed on acid-free paper

© Springer-Verlag London Limited 2007

MATLAB® is a registered trademark of The MathWorks, Inc., 3 Apple Hill Drive, Natick, MA 01760-2098, USA. http://www.mathworks.com

Aspen Plus® is a registered trademark of Aspen Technology, Inc., Ten Canal Park, Cambridge, MA 02141-2201, USA. http://www.aspentech.com/

9 8 7 6 5 4 3 2 1

Springer Science+Business Media
springer.com

Advances in Industrial Control

Professor Emeritus O.P. Malik
Department of Electrical and Computer Engineering
University of Calgary
2500, University Drive, NW
Calgary
Alberta
T2N 1N4
Canada

Professor K.-F. Man
Electronic Engineering Department
City University of Hong Kong
Tat Chee Avenue
Kowloon
Hong Kong

Professor G. Olsson
Department of Industrial Electrical Engineering and Automation
Lund Institute of Technology
Box 118
S-221 00 Lund
Sweden

Professor A. Ray
Pennsylvania State University
Department of Mechanical Engineering
0329 Reber Building
University Park
PA 16802
USA

Professor D.E. Seborg
Chemical Engineering
3335 Engineering II
University of California Santa Barbara
Santa Barbara
CA 93106
USA

Doctor K.K. Tan
Department of Electrical Engineering
National University of Singapore
4 Engineering Drive 3
Singapore 117576

Professor Ikuo Yamamoto
The University of Kitakyushu
Departmentof Mechanical Systems and Environmental Engineering
Faculty of Environmental Engineering
1-1, Hibikino, Wakamatsu-ku
Kitakyushu, Fukuoka, 808-0135
Japan

To the memory of my mother, Professor Linxian Feng

Jie Bao

To my wife Janet and my son Geoffrey

Peter L. Lee

Series Editors' Foreword

The series *Advances in Industrial Control* aims to report and encourage technology transfer in control engineering. The rapid development of control technology has an impact on all areas of the control discipline. New theory, new controllers, actuators, sensors, new industrial processes, computer methods, new applications, new philosophies…, new challenges. Much of this development work resides in industrial reports, feasibility study papers and the reports of advanced collaborative projects. The series offers an opportunity for researchers to present an extended exposition of such new work in all aspects of industrial control for wider and rapid dissemination.

Seminal contributions to the theory of dissipative and passive systems date from the early 1970s. Since that time the concepts have been used to design controllers for robotic systems and more recently, the thrust by control system design has led to some new contributions to the literature.

A key concept in dissipative systems theory is that of a storage function. This may be considered to be an abstraction of the idea of a store of energy within the system. The second key concept is a supply rate function and this can be viewed as an abstraction of the supply rate of energy to a system. Although the storage function and the supply rate function are abstract ideas, their value lies in the fact that they can often be identified with physical energy quantities within a real system.

Thus, a dissipative system is one for which the increase in internal energy is no greater than the energy supplied to it so the storage function quantifies internal energy or stored energy and the supply rate function prescribes the rate of energy supplied to the system. A passive system is then a dissipative system having a particular form of supply rate function, namely one expressed as an inner product of system input and output vectors.

The power of these dissipative system quantities lies in their links with system stability results and their ability to analyse physical systems described by nonlinear and linear models. In the field of process control, B.E. Ydstie and colleagues have demonstrated deep but natural links between thermodynamical systems and the notions of dissipative and passive systems theory. This *Advances in Industrial*

Control monograph by Jie Bao and Peter Lee makes a major step-forward in the literature of the passive systems approach to process control.

The monograph opens with a chapter on the fundamental ideas of dissipative and passive systems which covers the basic definitions and system properties before moving on to define passivity indices, and the methods of input feed forward and output feedback to create passivity. A gravity-feed tank system and heat exchanger process are used to illustrate the fundamental results of the chapter.

The next chapters explore the implications of the passivity systems approach to four key process control topics: robust control design, decentralised control, fault-tolerant control design and process controllability analysis; one chapter is devoted to each topic. These are all topics from the mainstream of process control applications and new results and insights are found in all four chapters of the volume. Passivity-based robust control designs are compared with H_∞ control designs in Chapter 3. Some of the decentralized and block-decentralised passivity-based robust controllers designed in Chapter 4 use PI controllers reflecting the widespread industrial use of these controllers. The fault-tolerant control designs of Chapter 5 attempts to achieve fault tolerance whilst avoiding or minimising controller redundancy. Finally, considerable academic and industrial interest in integrated process and controller design methods has motivated the use of passivity concepts in assessing process controllability; this forward-looking material is presented in Chapter 6.

The last chapter of the monograph, Chapter 7, is authored by Katalin Hangos and Gábor Szederkényi and this chapter delves into the fundamental links between the notions of thermodynamics and the constructs of passivity systems theory. It is an illuminating chapter that also considers the ideas of Hamiltonian system models and how models of process industry systems are constructed.

This excellent entry to the *Advances in Industrial Control* series contains new theoretical and applications-related results. It collects together the recent work of the authors to permit a cohesive presentation of passivity system theory as applied to process control. It is useful to note the wide range of industrial process models used in the examples. These include heat exchangers, a continuous stirred-tank reactor, distillation column, supercritical fluid extraction process, boiler furnace control, high-purity distillation column along with some purely academic examples. Many of these process models are standard in the process control literature and this facilitates comparisons of the results of the new methods with those already found in the published literature.

The potential readership for the monograph includes engineers and researchers from the process industries who may wish to exploit the methods and results directly. Research students on Masters and doctoral programmes in process and individual control will find the monograph an inspiring and interesting addition to their research literature. The wider readership of the control engineering and academic community may well find knowledge in this monograph that will transfer to other application fields; consequently the monograph is a very welcome new addition to the *Advances in Industrial Control* series.

M.J. Grimble and M.A. Johnson
Glasgow, Scotland, U.K.

Preface

Passive systems are intuitively appealing. Such systems do not "generate" energy internally, and hence are easier to control and to guarantee that the controlled response is stable. An understanding of the conditions that govern when and how any given system may be passive is thus an important approach in designing control systems. It is only in recent times that interest in using such approaches in the process industries has emerged.

This book is the first attempt to address passivity-based developments systematically in process control. It is written for a wide readership, including the industrial, engineering and academic communities. We have made an effort to present the theory backed by intuitive explanations, illustrative examples and/or case studies in all main chapters. The MATLAB® routines and controller parameters for all examples as well as a library of functions that implement the system analysis and control design methods developed in this book are available at http://www.springer.com/978-1-84628-892-0.

We have assumed that the readers have a working knowledge of engineering mathematics and that they have had some exposure to linear control theory. Some more advanced mathematical tools are introduced when necessary. This book presents the reader with both the conceptual framework and practical tools for passivity-based system analysis and control.

The authors are grateful for the contribution of Professor Katalin M. Hangos and Dr Gábor Szederkényi of the Systems and Control Laboratory, Computer and Automation Research Institute, Hungarian Academy of Sciences, who have written Chapter 7 of this book. This chapter makes clear the link between thermodynamics, Hamiltonian systems and passivity and how this linkage can be exploited in the design of passivity-based control systems.

This book is largely based on our recent research results. We wish to thank our co-workers and students, Dr Osvaldo Rojas, Dr Wenzhen Zhang, Dr Steven W. Su, Mr Kwong Ho Chan and Mr Herry Santoso for their contributions on the projects related to the subject of the present book. Dr Osvaldo Rojas also helped in proofreading some of the chapters.

We wish to express our gratitude to Professor Michael Johnson for his inspiration to prepare the book. Most importantly, we would like to thank our wives, who have continued to support us through the long hours that any such effort requires.

The University of New South Wales, Australia   *Jie Bao*
University of South Australia, Australia   *Peter L. Lee*
February 2007

Contents

1   **Introduction** 1

2   **Dissipativity and Passivity** 5
    2.1   Concept of Passive Systems 5
    2.2   Properties of Passive Systems 11
       2.2.1   Stability of Passive Systems 11
       2.2.2   Kalman–Yacubovich–Popov Property 12
       2.2.3   Input-Output Property 14
       2.2.4   Phase-related Properties 17
    2.3   Interconnection of Passive Systems 21
    2.4   Passivity Indices 24
       2.4.1   Excess and Shortage of Passivity 24
       2.4.2   Passivity Indices for Linear Systems 28
    2.5   Passivation 29
       2.5.1   Input Feedforward Passivation 29
       2.5.2   Output Feedback Passivation 30
    2.6   Passivity Theorem 32
    2.7   Heat Exchanger Example 36
    2.8   Summary 41

3   **Passivity-based Robust Control** 43
    3.1   Introduction 43
       3.1.1   Uncertainties 44
       3.1.2   Robust Stability 46
    3.2   Characterization of Uncertainties 47
       3.2.1   Uncertainty Bound Based on IFP 47
       3.2.2   Uncertainty Bounds Based on Simultaneous IFP and
           OFP 51
    3.3   Passivity-based Robust Control Framework 56
       3.3.1   Robust Stability Condition 56
       3.3.2   Robust Stability and Nominal Performance 57

        3.3.3   Advantages and Limitations of Passivity-based Robust
                Control 59
        3.3.4   Robust Control Design 59
        3.3.5   Example of CSTR Control........................... 65
    3.4 Combining Passivity with the Small Gain Condition 69
        3.4.1   Robust Stability Condition Based on Passivity and Gain 70
        3.4.2   Control Synthesis 73
        3.4.3   Robust Control of a Mixing System.................. 77
    3.5 Passive Controller Design 80
        3.5.1   Problem Formulation 83
        3.5.2   Contraction Map 84
        3.5.3   Synthesis of SPR/$\mathcal{H}_\infty$ Control 84
        3.5.4   Control Design Procedure 85
        3.5.5   Illustrative Example 86
    3.6 Summary... 87

4   **Passivity-based Decentralized Control** 89
    4.1 Introduction 89
    4.2 Decentralized Integral Controllability 91
        4.2.1   Passivity-based DIC Condition 93
        4.2.2   Computational Methods.............................. 94
    4.3 DIC Analysis for Nonlinear Processes 97
        4.3.1   DIC for Nonlinear Systems 97
        4.3.2   Sufficient DIC Condition for Nonlinear Processes 98
        4.3.3   Computational Method for Nonlinear DIC Analysis 100
        4.3.4   Nonlinear DIC Analysis for a Dual Tank System 101
    4.4 Block Decentralized Integral Controllability 103
        4.4.1   Conditions for BDIC............................... 105
        4.4.2   Pairing Based on BDIC 107
        4.4.3   BDIC Analysis of the SFE Process 108
    4.5 Dynamic Interaction Measure 110
        4.5.1   Representing Dynamic Interactions 110
        4.5.2   Passivity-based Interaction Measure 112
        4.5.3   Examples 117
    4.6 Decentralized Control Based on Passivity 120
        4.6.1   Problem Formulation 120
        4.6.2   Decentralized Control of Boiler Furnace 121
    4.7 Summary... 122

5   **Passivity-based Fault-tolerant Control** 125
    5.1 Introduction 125
    5.2 Representation of Sensor/Actuator Faults 126
    5.3 Decentralized Unconditional Stability Condition 128
        5.3.1   Passivity-based DUS Condition 128
        5.3.2   Diagonal Scaling 130

	5.3.3	Achievable Control Performance	131
	5.3.4	Pairing for Dynamic Performance	132
5.4	Fault-tolerant Control Design for Stable Processes		135
	5.4.1	Fault-tolerant PI Control	135
	5.4.2	Decentralized Fault-tolerant $\mathcal{H}_2$ Control Design	139
	5.4.3	Selecting the Weighting Function $w(s)$	140
	5.4.4	Control Synthesis	141
	5.4.5	Illustrative Example	146
5.5	Fault-tolerant Control Design for Unstable Processes		149
	5.5.1	Static Output Feedback Stabilization	150
	5.5.2	Fault-tolerant Control Synthesis	152
	5.5.3	Illustrative Example	153
5.6	Hybrid Active-Passive Fault-tolerant Control Approach		156
	5.6.1	Failure Mode and Effects Analysis	156
	5.6.2	Fault Detection and Accommodation	157
	5.6.3	Control Framework	159
5.7	Summary		160

6 Process Controllability Analysis Based on Passivity 161
6.1	Introduction		161
6.2	Analysis Based on Extended Internal Model Control		163
	6.2.1	Extended Internal Model Control Framework	163
	6.2.2	Controllability Analysis for Stable Linear Processes	166
6.3	Regions of Steady-state Attainability		171
	6.3.1	Steady-state Region of Attraction	172
	6.3.2	Steady-state Output Space Achievable via Linear Feedback Control	178
	6.3.3	Steady-state Attainability by Nonlinear Control	181
	6.3.4	Numerical Procedure	184
	6.3.5	Case Study of a High-purity Distillation Column	185
6.4	Dynamic Controllability Analysis for Nonlinear Processes		187
6.5	Summary and Discussion		191

7 Process Control Based on Physically Inherent Passivity
by K.M. Hangos and G. Szederkényi 193
7.1	Thermodynamic Variables and the Laws of Thermodynamics		194
	7.1.1	Extensive Variables, Entropy, Intensive Variables	194
	7.1.2	Laws of Thermodynamics	197
	7.1.3	Nonequilibrium Thermodynamics	198
7.2	The Structure of State Equations of Process Systems		199
	7.2.1	State Variables and Order of Systems	200
	7.2.2	Conservation Balances and Mechanisms	201
	7.2.3	Constitutive Equations	202
	7.2.4	State Equations of Process Systems	203
	7.2.5	Implications in Process Control	204

        7.2.6   Heat Exchanger Example 205
    7.3   Physically Motivated Supply Rates and Storage Functions 207
        7.3.1   Entropy-based Storage Functions 207
        7.3.2   Possible Choices of Inputs and Outputs 210
        7.3.3   Storage Function of the Heat Exchanger Example 210
    7.4   Hamiltonian Process Models 212
        7.4.1   System Structure and Variables 212
        7.4.2   Generalized Hamiltonian Systems 213
        7.4.3   Generalized Hamiltonian Systems with Dissipation 214
        7.4.4   Hamiltonian Description of the Heat Exchanger
                Example 215
    7.5   Case Study: Reaction Kinetic Systems 216
        7.5.1   System Description, Thermodynamic Variables and
                State-space Model 217
        7.5.2   The Reaction Simplex and the Structure of
                Equilibrium Points 217
        7.5.3   Physically Motivated Storage Function 218
        7.5.4   Passive Input-output Structure 219
        7.5.5   Local Hamiltonian Description of Reversible Reaction
                Networks 221
    7.6   Summary 224

A   Detailed Control Design Algorithms 225
    A.1   Solution to the BMI Problem in SPR/$\mathcal{H}_\infty$ Control Design 225
    A.2   DUS $\mathcal{H}_2$ Control Synthesis 226
        A.2.1   Final LMI 226
        A.2.2   SSDP Procedure 228

B   Mathematical Proofs 231
    B.1   Phase Condition for MIMO Systems 231
    B.2   Proof of Theorem 4.4 232
    B.3   Proof of Theorem 4.8 235
    B.4   Region of Steady-state Attainability 237
        B.4.1   Nominal Stability of Nonlinear IMC 237
        B.4.2   Proof of Theorem 6.6 239
        B.4.3   Positive Invariance of Region of Attraction 240
        B.4.4   Proof of Theorem 6.10 240

References 243

Index 251

Notation

Abbreviations

AS	asymptotically stable
BDIC	block decentralized integral controllable
BMI	bilinear matrix inequality
BRG	block relative gain
DCLI	decentralized closed-loop integrity
DIC	decentralized integral controllable
DUS	decentralized unconditional stability
ESPR	extended strictly positive real
GAS	globally asymptotically stable
GES	globally exponentially stable
IFP	input feedforward passivity
IMC	internal model control
ISE	integral square error
ITAE	integral time-weighted absolute error
LES	locally exponentially stable
LHP	left half plane
LMI	linear matrix inequality
LQG	linear quadratic Gaussian
LTI	linear time invariant
MIMO	multi-input multi-output
MP	minimum phase
NI	Niederlinski index
NMP	nonminimum phase
OFP	output feedback passivity
PID	proportional-integral-derivative
PR	positive real
RGA	relative gain array
RHP	right half plane
SDP	semidefinite programming

SISO	single-input single-output
SPR	strictly positive real
SSDP	successive semidefinite programming
SVD	singular value decomposition
ZSD	zero state detectable
ZSO	zero state observable

Symbols

A^T	transpose of matrix A
A^{-1}	inverse of matrix A
A^{-T}	transpose of inverse of matrix A
$A > 0$	matrix A is positive definite
$A \geq 0$	matrix A is positive semidefinite
A^*	complex conjugate transpose of complex matrix A
$\arg(c)$	angle of complex number c
$\mathbb{C}$	field of complex numbers
$\mathbb{C}^{m \times n}$	field of complex matrices of dimension $m \times n$
C^0 function	a function which is continuous
C^n function	a function which can be differentiated n times $(n > 1)$, leaving a continuous nth derivative
$\det(A)$	determinant of matrix A
$\dim(A)$	dimension of matrix A
$\mathcal{F}_l$	lower linear fractional transformation
$f_T(t)$	truncation operator of function $f(t)$
$\mathrm{Im}(A)$	imaginary part of complex matrix A
$\mathrm{In}(A)$	inertia of matrix A
$H : u \mapsto y$	mapping H from u to y
I	identity matrix
$\mathcal{L}_2^m$	$\mathcal{L}_2$ space with dimension m
$\mathcal{L}_{2e}^m$	extended $\mathcal{L}_2$ space with dimension m
$\mathbb{R}$	field of real numbers
$\mathbb{R}^{m \times n}$	field of real matrices of dimension $m \times n$
$\mathrm{Re}(A)$	real part of complex matrix A
$\sup$	supremum (the least upper bound)
$\mathrm{Tr}(A)$	trace of matrix A
$\Delta(s)$	uncertainty system
$\Delta_A(s)$	additive uncertainty
$\Delta_M(s)$	multiplicative uncertainty
$\bar{\sigma}(A)$	maximum singular value of matrix A
$\underline{\sigma}(A)$	minimum singular value of matrix A
$\sigma_i(A)$	ith singular value of matrix A
$\bar{\lambda}(A)$	maximum eigenvalue of matrix A

$\underline{\lambda}(A)$	minimum eigenvalue of matrix A
$\lambda_i(A)$	ith eigenvalue of matrix A
$\Lambda(A)$	RGA matrix of matrix A
$\lambda_{ij}(A)$	the ijth element of $\Lambda(A)$
$\Lambda_i(G(0))$	block relative gain (BRG) of the ith block $G_{ii}(0)$
$\mu(A)$	structured singular value of matrix A
ν	IFP index
ν_F	frequency-dependent IFP index
ν_{FB-}	passivity-based uncertainty measure in a frequency band
ν_I	passivity-based interaction measure
ν_{IA}	passivity-based interaction measure (additive uncertainty)
ν_{IM}	passivity-based interaction measure (multiplicative uncertainty)
ν_{S-}	sector bounded IFP measure (shortage of IFP)
ν_{sIA}	sector-based interaction measure (additive uncertainty)
ν_{sIM}	sector-based interaction measure (multiplicative uncertainty)
$\nu_-(G(s),\omega)$	shortage of IFP of $G(s)$ at frequency ω
ω	frequency
$:=$	state-space realization of a transfer function
$\triangleq$	defined as
$\|f\|_2$	2-norm of a function $f(t)$
$\|f\|_\infty$	∞-norm of a function $f(t)$
$\|f\|_{\infty-2}$	$\infty - 2$ norm of a function $f(t)$
$\|G\|_2$	2-norm of a system $G(s)$
$\|G\|_g$	generalized $\mathcal{H}_2$-norm of a system $G(s)$
$\|G\|_\infty$	$\mathcal{H}_\infty$-norm of a system $G(s)$
$< f,g >$	inner production of f and g
$\forall$	for all

1

Introduction

Passivity-based process control introduces an emerging area in process control – control design and system analysis based on the concept of passive systems. This monograph presents in a systematic approach the recent developments in robust, decentralized, and fault-tolerant process control, as well as process controllability analysis and nonlinear process control.

Passive systems are a class of processes that dissipate certain types of physical or virtual energy, described by Lyapunov-like functions. Passivity theory has been one of the cornerstones of nonlinear control theory since the 1970s. However, its application in process control has not been seen until recently. Defined as an input output property of process systems, the concept of passivity is particularly useful in stability analysis for interconnected systems. For example, a strictly passive system with a negative feedback of a passive system is stable (subject to the zero-state detectability condition). For a given system, its excess or shortage of passivity (which is called the passivity index) can be quantified by the feedback and feedforward required to render it passive. The shortage of passivity of a system (for example, a process) can be compensated for by the excess of passivity of another system (for example, a controller) to maintain closed-loop stability. This motivates control design based on passivity. The theoretical foundation of passive systems is introduced in Chapter 2.

The concept of passive systems is used in the development of robust process control. Robustness is an important issue in process control because uncertainties in process models are inevitable and could be significant in many cases. One can design a robust controller with excessive passivity that compensates for the "worst case" shortage of passivity of the process in the presence of a model–plant mismatch. This leads to characterization of process uncertainties in terms of their passivity and corresponding control synthesis methods. These issues are explored in Chapter 3 of this book.

Chapter 4 discusses a passivity-based decentralized control approach. Decentralized control is widely used in the process industry. This includes fully decentralized (multiloop) and block diagonal (multi-unit) control systems.

The passivity-based conditions are useful in decentralized control because they can be used to determine the stability of interconnected systems of complex process systems and decentralized control blocks, according to their passivity indices and the way they connect to each other. Passivity-based decentralized control can be less conservative than conventional approaches based on generalized diagonal dominance because it takes into account not only how large the interactions are but also how the subsystems interact with each other. The developments include interaction analysis, control structure selection (*e.g.*, pairing) and control system design.

In process control applications, failures of control components such as actuators, sensors or controllers are often encountered. These problems degrade the performance of the control system and also may induce instability, which could cause serious safety problems. With the increasing reliance on automatic control systems, fault-tolerant control becomes an important issue in the process industries. At present, most fault-tolerant control systems are based on the techniques of having redundancy in key controllers. A backup controller is employed once the failure of the main controller is detected. However, the control loop failure may not be detected swiftly and accurately. It also requires a significant number of redundant control components, which may increase the system cost to an unacceptable level. Based on the passivity conditions, a decentralized fault-tolerant approach that requires zero or very low level redundancy is developed. The basic idea is simple: a strictly passive multivariable plant can be stabilized by any decentralized passive controller. The decentralized passive controller remains passive when one or more of its subloops are arbitrarily detuned or taken out of service. This can be extended to nonpassive processes where a decentralized controller with excessive passivity is needed. Combining with existing fault detection and accommodation techniques, the passivity-based approach can be used to develop fault-tolerant control systems with minimum redundancy. Chapter 5 provides a description of fault-tolerant control, its basic framework, extensions and integration with existing fault detection and accommodation techniques.

Control systems are playing an increasingly important role in manufacturing industries and have become an indispensable part of any process plant. Therefore, it is very important to ensure that the outcome of process design can be easily controlled by feedback control systems to achieve effective disturbance rejection (for reduced product variability) and reference tracking (for fast and smooth transitions from one operating condition to another). Process controllability can be quantitatively measured by the best achievable dynamic control performance. A controllability measure which can be used in the early stages of process design will be very useful. Passive systems are minimum phase and thus very easy to control via output feedback, even if they are highly nonlinear and/or coupled. Such controllers include any multiloop PI/PID controllers with positive controller gain. Therefore, the controllability of a process can be inferred from its degree of passivity. Most chemical plants consist of multiple process units with recycle and bypass streams, which often make the controllability analysis very difficult. This chapter presents a pragmatic approach to controllability analysis based on open-loop process models and can be used in the early stages of process design.

Recent research has shown the very interesting link between passivity and thermodynamics. It is possible to determine the inherent passivity of a process from its physical properties. Chapter 7 presents recent work in revealing the connections between thermodynamics, Hamiltonian systems and passivity, and the applications in stabilization of nonlinear process systems.

2

Dissipativity and Passivity

This chapter introduces the concepts of passive and dissipative systems which lay the foundation of the developments described in this book. Most of the definitions follow Willems [138, 139], Byrnes *et al.* [24] and Sepulchre *et al.* [110] with possibly different notations. The implication of passivity is discussed in terms of the input-output behavior and stability of the process system.

2.1 Concept of Passive Systems

Much of the discussion presented in this chapter is related to system stability. Therefore, we start with a brief review of the stability of nonlinear systems. Consider a nonlinear system:

$$\frac{\mathrm{d}x}{\mathrm{d}t} = f(x, u), \tag{2.1}$$

where $x \in X \subset \mathbb{R}^n$ and $u \in U \subset \mathbb{R}^m$ are the state and input vector variables, respectively. The stability of this system is concerned with its *free* dynamics when the input variable $u = 0$. Assume that

$$f^*(x) = f(x, 0), \tag{2.2}$$

where the components of the n dimensional vector $f^*(x)$ are local Lipschitz functions of x, *i.e.*, $f^*(x)$ satisfies the following Lipschitz condition:

$$\|f^*(x_1) - f^*(x_2)\| \le L \|x_1 - x_2\| \tag{2.3}$$

for all x_1, x_2 in a neighbourhood of x_0, where L is a positive constant and $\|\cdot\|$ is the Euclidean norm (*i.e.*, $\|x\| = \sqrt{x^T x}$). The Lipschitz condition guarantees that

$$\frac{\mathrm{d}x}{\mathrm{d}t} = f^*(x) \tag{2.4}$$

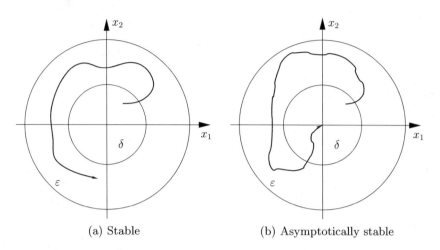

(a) Stable   (b) Asymptotically stable

Fig. 2.1. Lyapunov stability

has a unique solution with the initial condition $x(0) = x_0$. A point $x^* \in X$ is called an equilibrium point of (2.4) if $f^*(x^*) = 0$. The equilibrium point $x = 0$ is *stable* if for each $\varepsilon > 0$, $\delta = \delta(\varepsilon) > 0$ such that $\|x(0)\| < \delta$ implies that $\|x(t)\| < \varepsilon$ for all $t \geq 0$ (as shown in Figure 2.1a for $X \subset \mathbb{R}^2$). This equilibrium point is said to be *asymptotically stable* (AS) if it is stable and δ can be chosen such that $\|x(0)\| < \delta$ implies that $x(t)$ approaches the origin as t tends to infinity (as shown in Figure 2.1b). When the origin is asymptotically stable, the *region of attraction* is defined as the set of initial points $x(0)$ such that the solution of (2.4) approaches the origin as t tends to infinity. If the region of attraction is the entire state-space X, then the origin is *globally asymptotically stable* (GAS) [66]. Unlike linear systems, a nonlinear system may have multiple equilibrium points, of which some are stable and some are unstable. A sufficient condition for the stability of an equilibrium point is given by the Lyapunov stability criterion, which can be used to determine the stability of an equilibrium point without solving the state equation. Let $V(x)$ be a continuously differentiable (also denoted as C^1) scalar function defined in X that contains the origin. A function $V(x)$ is said to be *positive definite* if

$$V(0) = 0 \quad \text{and} \quad V(x) > 0, \quad \forall\, x \neq 0. \tag{2.5}$$

It is said to be *positive semidefinite* if

$$V(x) \geq 0, \quad \forall\, x. \tag{2.6}$$

Similarly, a function $V(x)$ is said to be *negative definite* if $V(0) = 0$ and $V(x) < 0$ for $x \neq 0$ and is said to be *negative semidefinite* if $V(x) \leq 0$ for all x.

Theorem 2.1 (Lyapunov stability criterion [67]). *Let $x = 0$ be an equilibrium point of a system described by (2.4). Function f^* is locally Lipschitz* and X contains the origin. The origin is stable if there exists a C^1 positive definite function $V(x) : X \to \mathbb{R}$ such that $\frac{\mathrm{d}V(x)}{\mathrm{d}t}$ is negative semidefinite and it is asymptotically stable if $\frac{\mathrm{d}V(x)}{\mathrm{d}t}$ is negative definite, where $\frac{\mathrm{d}V(x)}{\mathrm{d}t}$ is the derivative along the trajectory of (2.4), i.e.,

$$\frac{\mathrm{d}V(x)}{\mathrm{d}t} = \frac{\partial V(x)}{\partial x} f^*(x). \tag{2.7}$$

The function $V(x)$ in the above theorem, if it exists, is called a *Lyapunov function.*

A stronger type of stability is called *exponential stability*, which is defined as follows:

Definition 2.2 (Exponential stability [67]). *A system is globally exponentially stable (GES) if and only if there exists a Lyapunov function $V(x)$ such that*

$$\rho_1|x|^2 \leq V(x) \leq \rho_2|x|^2, \tag{2.8}$$

and with zero input,

$$\frac{\mathrm{d}V(x(t))}{\mathrm{d}t} \leq -\rho_3|x|^2, \tag{2.9}$$

where $\rho_i > 0$, $i = 1, 2, 3$ are suitable scalar constants. If these conditions hold, it follows that there exists some constant $\rho \geq 0$ such that with $x(0) = x_0$,

$$|x(t)| \leq \rho|x_0|e^{-\rho_3 t/2} \;\; \forall \, t \geq 0. \tag{2.10}$$

If the above condition is valid for x only in a neighbourhood of $x = 0$, the system is locally exponentially stable (LES).

First introduced by Popov [95], the concept of passive systems originally arose in the context of electrical circuit theory. A network consisting of only passive components, *e.g.*, inductors, resistors and capacitors, does not generate any energy and therefore is stable (*e.g.*, [6, 49]). In the early 1970s, Willems [138, 139] developed a systematic framework for dissipative systems, including passive systems, by introducing the notation of a storage function and a supply rate. Passivity, dissipativity and relevant stability conditions are cornerstones of modern control theory. In this section, an introduction to passive systems is presented through a very simple example of a gravity tank, followed by rigorous definitions.

Example 2.3 (Gravity tank). Consider the gravity under flow tank system illustrated in Figure 2.2. Assume that the input is the inlet volumetric flow rate $u = F_i(t)$, the state variable is the liquid level $x(t)$ and the output variable is the liquid pressure $y = p(t) = \rho g x(t)$. (Liquid pressure measurement is often used in level control.) Suppose that the outlet is flowing under the influence of gravity, *i.e.*,

$$F_o(t) = C_v \sqrt{x(t)}, \tag{2.11}$$

where C_v denotes the valve coefficient and F_o is the mass flow rate. The mass balance is given by

$$\rho A \frac{dx(t)}{dt} = \rho F_i(t) - \rho F_o(t) = \rho F_i(t) - \rho C_v \sqrt{x(t)}, \qquad (2.12)$$

leading to

$$\frac{dx(t)}{dt} = -\frac{C_v}{A} \sqrt{x(t)} + \frac{1}{A} F_i(t),$$
$$y(t) = p(t) = \rho g x(t), \qquad (2.13)$$

where A is the cross-sectional area of the tank and ρ is the density of the liquid. Denote the mass in the tank as m. Half of the potential energy stored in the tank is given by the following equation:

$$S(t) = S(x(t)) = \frac{1}{2} m(t) g x(t) = \frac{1}{2} [\rho A x(t)] g x(t) = \frac{1}{2} A \rho g x^2(t). \quad (2.14)$$

The inlet flow into the system increases the potential energy in the tank. The increment of potential energy per unit time can be represented by a function of the input and output:

$$w(t) = y(t) u(t) = \rho g F_i(t) x(t). \qquad (2.15)$$

The rate of change of the potential energy is given by taking the derivative along the trajectory of $x(t)$:

$$\frac{dS(t)}{dt} = \frac{\partial S}{\partial x} \frac{dx}{dt} = A \rho g x(t) \left[\frac{1}{A} \left(F_i(t) - C_v \sqrt{x(t)} \right) \right] \qquad (2.16)$$

$$= -C_v \rho g x(t) \sqrt{x(t)} + \rho g F_i(t) x(t) \qquad (2.17)$$

$$< w(t). \qquad (2.18)$$

Note that in the range of definition of x, the first term of (2.17) is always negative. Therefore the rate of change of the stored energy in the tank is less than that supplied to it by the inlet flow rate (represented by $w(t)$). As such, the tank system "dissipates" its potential energy through both the inlet flow (F_i) and the liquid pressure p, which is a function of both the input and output. This is called a *dissipative system*. Because the potential energy $S(t)$ is a positive definite function of the state variable $x(t)$, it can be treated as a Lyapunov function. When $F_i(t) = 0$,

$$\frac{dS(t)}{dt} < 0, \quad \forall\, x \neq 0. \qquad (2.19)$$

Therefore, the equilibrium $x = 0$ is asymptotically stable (AS). If the outlet valve is completely shut off (*i.e.*, $C_v = 0$), then the energy flow into the tank is totally stored. In this case, this process becomes *lossless* and the equilibrium $x = 0$ is stable.

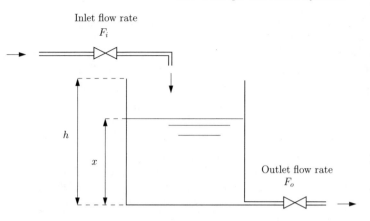

Fig. 2.2. A gravity tank system

Comparing (2.17) with (2.19), it can be seen that by introducing the energy function, (2.17) gives the stability of a free system (with zero input) and also how its input and output affect the state variable. If we generalize the energy function to any nonnegative function of the states, then we can define a class of nonlinear processes. Consider the following nonlinear system:

$$H : \begin{cases} \dot{x} & = f\left(x, u\right) \\ y & = h(x, u), \end{cases} \tag{2.20}$$

where $x \in X \subset \mathbb{R}^n$, $u \in U \subset \mathbb{R}^m$ and $y \in Y \subset \mathbb{R}^m$ are the state, input and output variables, respectively, and X, U and Y are state, input and output spaces, respectively. The representation $x(t) = \phi(t, t_0, x_0, u)$ is used to denote the state at time t reached from the initial state x_0 at t_0.

Definition 2.4 (Supply rate [138]). *The supply rate $w(t) = w(u(t), y(t))$ is a real valued function defined on $U \times Y$, such that for any $u\left(t\right) \in U$ and $x_0 \in X$ and $y(t) = h(\phi(t, t_0, x_0, u))$, $w(t)$ satisfies*

$$\int_{t_0}^{t_1} |w(t)|\, dt < \infty \tag{2.21}$$

for all $t_1 \geq t_0 \geq 0$.

Definition 2.5 (Dissipative systems [138]). *System H with supply rate $w(t)$ is said to be dissipative if there exists a nonnegative real function $S(x)$: $X \to \mathbb{R}^+$, called the storage function, such that, for all $t_1 \geq t_0 \geq 0$, $x_0 \in X$ and $u \in U$,*

$$S(x_1) - S(x_0) \leq \int_{t_0}^{t_1} w(t)dt, \tag{2.22}$$

where $x_1 = \phi(t_1, t_0, x_0, u)$ and $\mathbb{R}^+$ is a set of nonnegative real numbers.

The above condition states that a system is dissipative if the increase in its energy (storage function) during the interval (t_0, t_1) is no greater than the energy supplied (via the supply rate) to it. If the storage function is differentiable, *i.e.*, it is C^1, then we can write (2.22) as

$$\frac{\mathrm{d}S\left(x\left(t\right)\right)}{\mathrm{d}t} \leq w(t). \tag{2.23}$$

The interpretation is that the rate of increase of energy is no greater than the input power.

According to the above definition, a storage function has to be positive semidefinite. The next definition describes the notion of available storage, the largest amount of energy that can be extracted from the system given the initial condition $x\left(0\right) = x$:

Definition 2.6 (Available storage [138]). *The available storage, S_a of a system H with supply rate w, is the function $S_a : X \rightarrow \mathbb{R}^+$ defined by:*

$$S_a\left(x\right) \triangleq \sup_{\substack{x(0)=x \\ u(t)\in U \\ t_1>0}} \left\{ -\int_0^{t_1} w\left(u\left(t\right), y\left(t\right)\right) dt \right\}. \tag{2.24}$$

The available storage is nonnegative, since $S_a\left(x\right)$ is the supremum over a set of values including the zero element. The available storage function plays an important role in dissipative/passive systems. If a system is dissipative, the available storage function $S_a(x)$ is finite for each $x \in X$. Moreover, any possible storage function $S(x)$ satisfies

$$0 \leqslant S_a(x) \leqslant S(x) \tag{2.25}$$

for each $x \in X$. If S_a is a continuous (C^0) function, then S_a itself is a possible storage function. Conversely, if $S_a\left(x\right)$ is finite for every $x \in X$, then the system is dissipative with respect to the supply rate $w(t)$.

The supply rate can be any function defined on the input and output space that satisfies (2.21). When a bilinear supply rate is adopted, passive systems can be defined as:

Definition 2.7 (Passive systems [24]). *A system is said to be passive if it is dissipative with respect to the following supply rate:*

$$w\left(u\left(t\right), y\left(t\right)\right) = u^T\left(t\right) y\left(t\right), \tag{2.26}$$

and the storage function $S\left(x\right)$ satisfies $S(0) = 0$.

Two extreme cases of passive systems are lossless and state strictly passive systems:

Definition 2.8 (Lossless systems [24]). *A passive system* H *with storage function* $S(x)$ *is said to be lossless if for all* $t_1 \geq t_0 \geq 0$, $x_0 \in X$ *and* $u \in U$,

$$S(x) - S(x_0) = \int_{t_0}^{t_1} y^T(t) u(t) \, dt. \tag{2.27}$$

Definition 2.9 (State strictly passive systems [24]). *A passive system* H *with storage function* $S(x)$ *is said to be state strictly passive if there exists a positive definite function* $V : X \rightarrow \mathbb{R}^+$ *such that for all* $t_1 \geq t_0 \geq 0$, $x_0 \in X$ *and* $u \in U$,

$$S(x) - S(x_0) = \int_{t_0}^{t_1} y^T(t) u(t) \, dt - \int_{t_0}^{t_1} V(x(t)) \, dt. \tag{2.28}$$

This definition is referred to as *strict passivity* in [24]. Here we define it as *state strict passivity* to discriminate it from other types of strict passivity discussed later in this book, such as strict input passivity and strict output passivity.

In the tank system example, the storage function is the total potential energy stored in the tank system, given by (2.14). The supply rate given by (2.15) is the inner product of the input and output. Therefore, the tank system is state strictly passive when the outlet valve is open and is lossless when the outlet valve is closed. Storage functions are not limited to physical energies. Any nonnegative real functions defined on state variables can be understood as a type of *abstract energy*, like the Lyapunov functions. They are potential candidates for the storage functions. For example, for the tank system, if we choose the output as the liquid level x, then the supply rate $w(t) = F_i(t) x(t)$. With the storage function $S(x) = \frac{1}{2} A x^2$, it is obvious that

$$\frac{dS(x)}{dt} = -C_v x(t) \sqrt{x(t)} + F_i(t) x(t) < w(t), \tag{2.29}$$

which shows that the process is passive (more precisely, state strictly passive). In this case, the physical meanings of the storage function and supply rate are not explicit, making it more difficult to determine its passivity directly from our understanding of the mass and energy balance. However, the process possesses all the useful properties of passive systems that we are going to discuss in the next section.

2.2 Properties of Passive Systems

2.2.1 Stability of Passive Systems

In the tank example, we can see that the concept of passivity implies stability if a positive definite storage function is used. Because the storage function is only required to be positive semidefinite in Definition 2.5, stability is not always ensured by passivity. For example, if a system has two states $x = [x_1, x_2]^T$ and the storage function is positive semidefinite, e.g., $S(x) = \frac{1}{2}x_1^2$, then passivity with this storage function does not imply the stability of x_2. In this case, additional conditions on zero-state detectability and observability are required:

Definition 2.10 (Zero-state observability and detectability [24]). *A system as given in (2.20) is zero-state observable (ZSO) if for any $x \in X$,*

$$y(t) = h(\phi(t, t_0, x, 0)) = 0, \quad \forall\, t \geq t_0 \geq 0 \quad \text{implies } x = 0, \qquad (2.30)$$

and the system is locally ZSO if there exists a neighbourhood X_n of 0, such that for all $x \in X_n$, (2.30) holds. The system is zero-state detectable (ZSD) if for any $x \in X$,

$$y(t) = h(\phi(t, t_0, x, 0)) = 0, \quad \forall\, t \geq t_0 \geq 0 \quad \text{implies } \lim_{t \to \infty} \phi(t, t_0, x, 0) = 0,$$
$$(2.31)$$

and the system is locally ZSD if there exists a neighbourhood X_n of 0, such that for all $x \in X_n$, (2.31) holds.

With the definition of zero-state detectability (ZSD), the link between passivity and Lyapunov stability can be established:

Theorem 2.11 (Passivity and stability [110]). *Let a system H (as represented in (2.20)) be passive with a C^1 storage function $S(x)$ and $h(x, u)$ be C^1 in u for all x. Then the following properties hold:*

1. *If $S(x)$ is positive definite, then the equilibrium $x = 0$ of H with $u = 0$ is Lyapunov stable.*
2. *If H is ZSD, then the equilibrium $x = 0$ of H with $u = 0$ is Lyapunov stable.*
3. *If in addition to either Condition 1 or Condition 2, $S(x)$ is radially unbounded (i.e., $S(x) \to \infty$ as $\|x\| \to \infty$), then the equilibrium $x = 0$ in the above conditions is globally stable (GS).*

It can be also found that if system H is state strictly passive with a positive definite storage function, then the equilibrium $x = 0$ with $u = 0$ is asymptotically stable. The boundedness of the storage function implies the boundedness of the state variables. However, passivity tells more than just stability. It relates the input and output to the storage function and thus defines a set of useful input-output properties, which are explained in the next section.

2.2.2 Kalman–Yacubovich–Popov Property

One of the most important properties of passive systems is related to the following definition:

Definition 2.12 (Kalman–Yacubovich–Popov property [24]). *Consider a control affine system without throughput (as a special case of the system in (2.20)):*

$$H : \begin{cases} \dot{x} &= f(x) + g(x)u \\ y &= h(x), \end{cases} \tag{2.32}$$

where $x \in X \subset \mathbb{R}^n$, $u \in U \subset \mathbb{R}^m$ and $y \in Y \subset \mathbb{R}^m$. It is said to have the Kalman–Yacubovitch–Popov (KYP) property if there exists a C^1 nonnegative function $S(x) : X \to \mathbb{R}^+$, with $S(0) = 0$ such that

$$L_f S(x) = \frac{\partial S(x)}{\partial x} f(x) \leq 0, \tag{2.33}$$

$$L_g S(x) = \frac{\partial S(x)}{\partial x} g(x) = h^T(x), \tag{2.34}$$

for each $x \in X$.

The term $L_f S(x) = \frac{\partial S(x)}{\partial x} f(x)$ is called the *Lie derivative*, which is defined as follows:

Definition 2.13 (Lie derivative). *Given a C^1 nonlinear scalar function $S(x) : \mathbb{R}^n \to \mathbb{R}$ and a vector function:*

$$f(x) = [f_1(x), f_2(x), \cdots, f_n(x)]^T \in \mathbb{R}^n \to \mathbb{R}^n, \tag{2.35}$$

on a common domain $X \subset \mathbb{R}^n$. The derivative of $S(x)$ along f is defined as

$$L_f S(x) = \frac{\partial S(x)}{\partial x} f(x) = \sum_{i=1}^{n} \frac{\partial S(x)}{\partial x_i} f_i(x). \tag{2.36}$$

The repeated Lie derivative is defined as

$$L_f^k S(x) = \frac{\partial \left(L_f^{k-1} S(x) \right)}{\partial x} f(x), \tag{2.37}$$

with $L_f^0 S(x) = S(x)$.

Proposition 2.14 ([57]). *A system H which has the KYP property is passive, with a storage function $S(x)$. Conversely, a passive system having a C^1 storage function has the KYP property.*

For the tank system in Example 2.3, $f(x) = -\frac{C_v}{A}\sqrt{x}$, $g(x) = \frac{1}{A}$ and $h(x) = \rho g x(t)$. With the storage function defined in (2.14), it is easy to verify that the tank system has the KYP property. Because the liquid level $x(t) \geq 0$,

$$L_f S(x) = -\rho g C_v x(t) \sqrt{x(t)} \leq 0, \tag{2.38}$$

$$L_g S(x) = A\rho g x(t) \frac{1}{A} = \rho g x(t) = y(t). \tag{2.39}$$

For a linear time invariant (LTI) system, there exists a quadratic storage function $S(x) = x^T P x$ (with a positive definite matrix P), leading to the following linear version of the KYP condition:

Proposition 2.15 ([139]). *Consider a stable LTI system:*[1]

$$\dot{x} = Ax + Bu$$
$$y = Cx + Du, \tag{2.40}$$

where $x \in \mathbb{R}^n$, $u \in \mathbb{R}^m$ and $y \in \mathbb{R}^m$. This system is passive if and only if there exist matrices $P, L \in \mathbb{R}^{n \times n}, Q \in \mathbb{R}^{m \times n}$ and $W \in \mathbb{R}^{m \times m}$ with $P > 0$, $L > 0$ (positive definite) such that

$$A^T P + P A = -Q^T Q - L,$$
$$B^T P - C = -W^T Q, \tag{2.41}$$
$$W^T W = D + D^T.$$

For systems with relative degree 0 (*i.e.*, $D \neq 0$), the above condition can be represented using a linear matrix inequality (LMI), which is often referred to as the positive-real lemma:

Lemma 2.16 (Positive-real Lemma [21]). *A stable LTI system given in (2.40) with $D \neq 0$ is passive if and only if there exists a positive definite matrix P such that:*

$$\begin{bmatrix} A^T P + P A & P B - C^T \\ B^T P - C & -D - D^T \end{bmatrix} < 0. \tag{2.42}$$

When $D = 0$, the above condition is reduced to

$$A^T P + P A < 0, \tag{2.43}$$
$$B^T P = C. \tag{2.44}$$

Equations (2.43) and (2.44) are the linear versions of (2.33) and (2.34), respectively.

2.2.3 Input-Output Property

Obviously, while (2.33) is related to the stability, (2.34) defines an input-output property. The input-output property of passive systems is called *positive realness*:

---

[1] In this book, the linear system given in (2.40) is said to be *stable* if $\mathrm{Re}[\lambda_i(A)] < 0$, $\forall\, i = 1, \ldots, n$. The system is actually *asymptotically stable* according to Section 2.1.

Definition 2.17 (Positive real systems [24]). *A system is said to be positive real if for all $t_1 \geq t_0 \geq 0$, $u \in U$,*

$$\int_{t_0}^{t_1} y^T(t)u(t)dt \geqslant 0, \tag{2.45}$$

whenever $x(t_0) = 0$.

The initial condition of the state variable $x_0 = x(t_0) = 0$ (consequently $S(x_0) = 0$) is assumed because positive realness is only an input-output property, which says nothing about the states. Clearly, passive systems are positive real. To tell whether a positive real system is passive, we need an additional *reachability* condition:

Definition 2.18 (Reachability and controllability [138]). *The state-space of a dynamic system H (as in (2.20)) is said to be reachable from x_{-1} if for any $x \in X$, there exists a $t_{-1} \leq 0$ and $u \in U$ such that*

$$x = \phi(0, t_{-1}, x_{-1}, u). \tag{2.46}$$

It is said to be controllable to x_1 if for any $x \in X$, there exists a $t_1 \geq 0$ and $u \in U$ such that

$$x_1 = \phi(t_1, 0, x, u). \tag{2.47}$$

A positive real system is passive if any state is reachable from the origin and S_a is at least continuous (C^0). A thorough treatment of passive systems from the perspective of input-output systems can be found in [32]. In the case of linear systems, *positive realness* and *passivity* are synonyms, provided that the system is detectable.

The input-output relationship is often more conveniently represented by system operators. The system operator is a mapping defined on signal spaces. For example, system H with the input and output signals $u(t)$ and $y(t)$ can be understood as a mapping from u to y (with certain initial conditions on the state variable $x(t)$). In this case, $H : u(t) \longmapsto y(t)$ is a system operator and the system output can be represented as

$$y(t) = Hu(t). \tag{2.48}$$

The mapping from $y(t)$ to $u(t)$ is referred to as the inverse of H, denoted as H^{-1}. For a vector signal function on time $f(t) = [f_1(t), \ldots, f_m(t)]^T$, where $f_i(t)$ are scalar functions and $t \geq 0$, the "size" of the signal can be quantified by using norms. Here we introduce the so-called 2-norm:

Definition 2.19 (2-norm of a signal). *The 2-norm of a vector time-domain signal $f(t) \in \mathbb{R}^m$ is defined as*

$$\|f\|_2 \triangleq \left[\sum_{i=1}^{m} \int_0^\infty f_i^2(t)\,dt\right]^{\frac{1}{2}} = \sqrt{\int_0^\infty f^T(t)f(t)\,dt}. \tag{2.49}$$

Define the inner product as follows:

$$\langle f, g \rangle \triangleq \int_0^\infty f(t)^T g(t) \, dt. \tag{2.50}$$

Then (2.49) can be written as

$$\|f\|_2 = \sqrt{\langle f, f \rangle}. \tag{2.51}$$

The set of vector functions of $f : \mathbb{R}^+ \to \mathbb{R}^m$ which have a bounded 2-norm, i.e.

$$\|f\|_2 < \infty, \tag{2.52}$$

are called the $\mathcal{L}_2^m$ space (the superscript indicates the dimension). This is a Hilbert space (a linear space with inner product). The $\mathcal{L}_2^m$ space can be extended to allow functions that are unbounded, when $t \to \infty$, by introducing the truncation operator:

Definition 2.20 (Truncation operator [130]). *Let $f : \mathbb{R}^+ \to \mathbb{R}^m$. Then for each $T \geq 0$, the function $f_T(t)$ is defined by*

$$f_T(t) = \begin{cases} f(t), & 0 \leq t < T \\ 0, & t \geq T, \end{cases} \tag{2.53}$$

and is called the truncation of f to the interval $[0, T]$.

The space that consists of all functions f such that $f_T(t) \in \mathcal{L}_2^m$ is called the extension of $\mathcal{L}_2^m$, denoted as $\mathcal{L}_{2e}^m$. Now the definition of input-output stability can be given as follows:

Definition 2.21 (Input-output stability [130]). *Let $H : \mathcal{L}_{2e}^m \to \mathcal{L}_{2e}^p$. System H is said to be $\mathcal{L}_2$ stable if $Hu \in \mathcal{L}_2^p$ for any $u \in \mathcal{L}_{2e}^m$.*

The mapping H is said to have finite $\mathcal{L}_2$ gain if there exist finite constants γ and b such that for all $T \geq 0$,

$$\|(Hu)_T\|_2 \leq \gamma \|u_T\|_2 + b, \ \forall \, u \in \mathcal{L}_{2e}^m. \tag{2.54}$$

Using the above definition, we can define passivity from the perspective of the input-output property:

Definition 2.22 ([130]). *Let $H : u \in \mathcal{L}_{2e}^m \mapsto y \in \mathcal{L}_{2e}^m$. Then system H is passive if there exists some constant β such that*

$$\langle Hu, u \rangle_T = \langle y, u \rangle_T \geq \beta, \ \forall \, u \in \mathcal{L}_{2e}^m, \ \forall \, T \geq 0. \tag{2.55}$$

The above inequality is equivalent to the positive real condition given in (2.45), with the assumption $t_0 = 0$. The introduction of constant β is due to the fact that $x(t_0) = 0$ is not assumed in (2.55). One possible case is $\beta = S(x(0))$, where $S(x)$ is the storage function.

Because both (2.45) and (2.55) are symmetrical in terms of u and y, the following proposition is obvious:

Proposition 2.23 ([116]). *Consider a positive real system H which maps u to y. Its inverse (denoted H^{-1}) which maps y to u is also positive real if it exists.*

For stable linear systems, the above input-output property can be defined on the transfer functions by introducing positive real transfer functions.

Definition 2.24 (Positive real transfer function [139]). *A transfer function $G(s)$ is positive real if*

- $G(s)$ *is analytic in* $\mathrm{Re}(s) > 0$*;*
- $G(j\omega) + G^*(j\omega) \geq 0$ *for any frequency ω that $j\omega$ is not a pole of $G(s)$. If there are poles $p_1, p_2, \ldots, p_q$ of $G(s)$ on the imaginary axis, they are non-repeated and the residue matrix at the poles $\lim\limits_{s \to p_i} (s - p_i)G(s)$ $(i = 1, \ldots, q)$ is Hermitian and positive semidefinite.*

Transfer function $G(s)$ is said to strictly positive real (SPR) if

- $G(s)$ *is analytic in* $\mathrm{Re}(s) \geq 0$*;*
- $G(j\omega) + G^*(j\omega) > 0 \ \forall \ \omega \in (-\infty, +\infty)$*.*

Furthermore $G(s)$ is said to be extended strictly positive real (ESPR) if it is SPR and $G(j\infty) + G^*(j\infty) > 0$ [123].

Here, $G^*(j\omega)$ is the complex conjugate transpose of $G(j\omega)$.

Theorem 2.25 ([139]). *A linear system as given in (2.40) is passive (or strictly passive) if and only if its transfer function $G(s) := C(sI - A)^{-1}B + D$ is positive real (or strictly positive real).*

The above theorem (together with Definition 2.24) forms an input-output version of the positive-real lemma in the frequency domain. The above theorem is often used as the definition of linear passive systems. According to Theorem 2.25, $G_1(s) = \frac{1}{s+1}$ is a strictly passive system and $G_2(s) = \frac{1}{s}$ is a passive system. It is worth pointing out that *any* PID controller

$$K(s) = k_c \left[1 + \frac{1}{\tau_I s} + \tau_D s\right], \quad k_c > 0, \tag{2.56}$$

is passive. So is any multiloop PID controller.

2.2.4 Phase-related Properties

The above input-output property implies another interesting characteristic of passive systems – they are *phase bounded*. This is very obvious for SISO passive systems, because the condition $G(j\omega) + G^*(j\omega) \geq 0$ is then reduced to $\mathrm{Re}(G(j\omega)) \geq 0$, which means that the real part of their frequency response is always nonnegative. This is what the term "positive real" originally referred to. Clearly, the phase shift of a stable SISO passive system in response to a sinusoidal input is always within $[-90°, 90°]$ and the phase shift of a SISO strictly passive system is always within $(-90°, 90°)$.

The above statement is also true for multi-input multi-output (MIMO) linear systems. Here we adopt the following phase definition for MIMO systems given by Postlethwaite *et al.*:

Definition 2.26 (Phase of MIMO LTI systems [96]). *Consider an MIMO LTI system with a transfer function $G(s) \in \mathbb{C}^{m \times m}$. Perform the polar decomposition on its frequency response:*

$$
\begin{aligned}
G(j\omega) &= X(j\omega)\Lambda(j\omega)V^*(j\omega) \\
&= [X(j\omega)V^*(j\omega)][V(j\omega)\Lambda(j\omega)V^*(j\omega)] = U(j\omega)H(j\omega),
\end{aligned}
\tag{2.57}
$$

where $\Lambda(j\omega)$ is an $m \times m$ diagonal, real and nonnegative matrix; $X(j\omega)$ and $V(j\omega)$ are unitary matrices. $U(j\omega) = X(j\omega)V^*(j\omega)$ is also a unitary matrix and $H = V(j\omega)\Lambda(j\omega)V^*(j\omega)$ is a Hermitian matrix. The phase of the system at frequency ω is defined as the principal arguments of the eigenvalues of $U(j\omega)$.

Theorem 2.27 (Phase condition for MIMO LTI strictly passive systems [13]). *Consider an MIMO LTI system with a transfer function $G(s) \in \mathbb{C}^{m \times m}$. If the system is strictly passive, then its phase shift lies in the open interval $(-90°, 90°)$ for any real ω.*

The proof of the above theorem is given in Section B.1. If the frequency response of a stable linear system has a phase shift within $[-90°, 90°]$ for all frequencies, this system also satisfies both of the following conditions:

1. it is minimum phase;
2. the difference between the degree of the denominator polynomial and the degree of the numerator polynomial (*i.e.*, the relative degree) is less than 2.

This can be illustrated by a simple SISO case. Consider a stable and minimum phase transfer function $G(s) = \frac{p(s)}{q(s)}$ with $G(0) > 0$, where the numerator polynomial $p(s)$ is of mth order and the denominator polynomial $q(s)$ is of nth order. Because $G(s)$ has only left half plane (LHP) zeros and poles at frequency $\omega = \infty$, the phase shift will be $90°(n - m)$. Therefore, for the system to be phase bounded by $[-90°, 0°]$ at all frequencies, it must satisfy $n - m < 2$. (A positive phase shift will occur when $G(0) < 0$.)

Phase is not defined for nonlinear systems. However, the above phase-related conditions can be extended to nonlinear systems. The relative degree can be understood as the number of times one has to differentiate the output to have the input explicitly appearing. Therefore, we can define the relative degree for nonlinear systems as follows:

Definition 2.28 (Relative degree [61]). *A SISO control affine nonlinear system*

$$\begin{aligned} \dot{x} &= f\left(x\right) + g\left(x\right) u \\ y &= h\left(x\right), \end{aligned}$$ (2.58)

is said to have relative degree r at point x_0 if

1. *$L_g L_f^k h\left(x\right) = 0$ for all x in a neighbourhood of x_0 and all $k < r-1$;*
2. *$L_g L_f^{r-1} h\left(x_0\right) \neq 0$,*
   where $L_f^k h\left(x\right)$ is the kth order Lie derivative of h along f.
   A multivariable nonlinear control affine system as in the following equation:

$$\begin{aligned} \dot{x} &= f\left(x\right) + \sum_{j=1}^{q} g_j\left(x\right) u_j, \\ y_i &= h_i\left(x\right), \quad i = 1, \dots, p, \end{aligned}$$ (2.59)

has a vector relative degree given by $\{r_1, r_2, \cdots, r_p\}$ at a point x_0 if

1. *$L_{g_i} L_f^k h_i\left(x\right) = 0, \quad i = 1, \dots, p, \; k = 0, \dots, r_i - 2$ for all x in a neighbourhood of x_0.*
2. *The characteristic matrix $C\left(x\right)$, given by*

$$C\left(x\right) = \begin{bmatrix} L_{g_1} L_f^{r_1-1} h_1\left(x\right) & L_{g_2} L_f^{r_1-1} h_1\left(x\right) & \cdots & L_{g_p} L_f^{r_1-1} h_1\left(x\right) \\ L_{g_1} L_f^{r_2-1} h_2\left(x\right) & L_{g_2} L_f^{r_2-1} h_2\left(x\right) & \cdots & L_{g_p} L_f^{r_2-1} h_2\left(x\right) \\ \vdots & \vdots & \ddots & \vdots \\ L_{g_1} L_f^{r_p-1} h_p\left(x\right) & L_{g_2} L_f^{r_p-1} h_p\left(x\right) & \cdots & L_{g_p} L_f^{r_p-1} h_p\left(x\right) \end{bmatrix}_{p \times p}$$

is nonsingular at x_0. The total relative degree is defined as $r = \sum_{i=1}^{p} r_i$.

For the linear SISO system $\dot{x} = Ax + Bu$, $y = Cx$, the relative degree is equal to the difference between the degree of the denominator polynomial and the degree of the numerator polynomial of the transfer function $H(s) = C(sI - A)^{-1} B$ of the system. To extend the concept of minimum phase systems to nonlinear systems, we need to look at the zero dynamics:

Definition 2.29 (Zero dynamics). *Consider the system in (2.32) with the constraint $y = 0$, i.e.,*

$$\begin{aligned} \dot{x} &= f\left(x\right) + g\left(x\right) u, \\ 0 &= h\left(x\right). \end{aligned}$$ (2.60)

The constrained system (2.60) is called the zero-output dynamics, or briefly, the zero dynamics.

If the matrix $L_g h\left(0\right) \triangleq \left.\frac{\partial h(x)}{\partial x} g\left(x\right)\right|_{x=0}$ of the system in (2.32) is nonsingular and the distribution spanned by the vector fields $g_1\left(x\right), \cdots, g_m\left(x\right)$ is involutive in a neighbourhood of $x = 0$, then there exists new local coordinates (z, y) under which the system can be represented as the so-called normal form:

$$\begin{aligned} \dot{z} &= q\left(z, y\right), \\ \dot{y} &= b\left(z, y\right) + a\left(z, y\right) u. \end{aligned} \qquad (2.61)$$

The zero dynamics of system (2.32) are given by

$$\dot{z} = q\left(z, 0\right). \qquad (2.62)$$

Denote $q\left(z, 0\right)$ by $f_0\left(z\right)$. Then, the function $q\left(z, y\right)$ can be expressed in the form

$$q\left(z, y\right) = f_0\left(z\right) + p\left(z, y\right) y, \qquad (2.63)$$

where $p\left(z, y\right)$ is a smooth function (see [24]).

Definition 2.30 (Minimum phase nonlinear systems [24]). *Consider the system in (2.32). Suppose that $L_g h\left(0\right)$ is nonsingular. Then the system is said to be:*

1. *minimum phase if its zero dynamics are asymptotically stable in a neighbourhood of $z = 0$;*
2. *weakly minimum phase if there exists a positive differentiable function $W\left(z\right)$ with $W\left(0\right) = 0$, such that*

$$\frac{\partial W\left(z\right)}{\partial z} f_0\left(z\right) \le 0 \qquad (2.64)$$

in a neighbourhood of $z = 0$.

Similarly, we can define globally minimum phase and globally weakly minimum phase if the normal form and minimum phase are global. Now we are in the position to study the phase-related properties of nonlinear passive systems.

Theorem 2.31 ([24]). *Consider system H given in (2.32). Assume that* rank $\{L_g h\left(x\right)\}$ *is constant in a neighbourhood of $x = 0$. If system H is passive with a C^2 storage function $S\left(x\right)$ which is positive definite, then*

1. *$L_g h\left(0\right)$ is nonsingular and H has relative degree $\{1, \cdots, 1\}$.*
2. *The zero dynamics of H exist locally at $x = 0$, and H is weakly minimum phase.*

Because system H in consideration does not have a feedthrough term, its relative degree could not be below $\{1, \cdots, 1\}$. A passive SISO nonlinear system has a relative degree of 1 or 0 (if there is a feedthrough term). The above theorem shows that nonlinear passive systems have phase-related input-output properties similar to those their linear counterparts possess. These properties imply output feedback stability conditions which will be discussed in the next section.

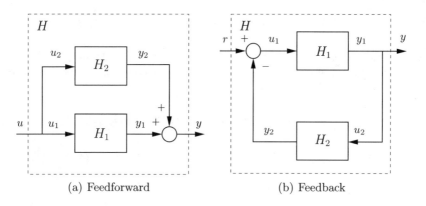

(a) Feedforward   (b) Feedback

Fig. 2.3. Interconnections of passive systems

2.3 Interconnection of Passive Systems

The phase-related properties of passive systems imply important output feedback stability conditions, which can be used to determine the stability of networks of interconnected systems. A passive system is very easy to control via output feedback. For example, a linear passive system ($e.g.,$ $G(s) = \frac{1}{s}$) can be stabilized by any proportional only controller with a positive gain. Similarly, we have the following stability condition for nonlinear systems:

Theorem 2.32. $For\ a\ nonlinear\ passive\ system\ H\ given\ in\ (2.32),\ a\ proportional\ only\ output\ feedback\ control\ law\ u = -ky\ asymptotically\ stabilizes\ the\ equilibrium\ x = 0\ for\ any\ k > 0,\ provided\ that\ H\ is\ ZSD.$

$Proof.$ Assume that H is passive with storage function $S(x)$. For $u = -y$, the time derivative of S satisfies

$$\dot{S}(x) \leq -ky^T y < 0, \quad \forall\ y \neq 0. \tag{2.65}$$

The bounded solution of $\dot{x} = f(x, -y)$ is confined in $\{x|\ h(x) = 0\}$. If H is ZSD, then $x \to 0$.

The output feedback stability condition is not limited to static feedback:

Theorem 2.33 (Interconnections of passive systems). $Suppose\ that\ systems\ H_1\ and\ H_2\ are\ passive\ (as\ shown\ in\ Figure\ 2.3).\ Then\ the\ two\ systems,\ one\ obtained\ by\ the\ parallel\ interconnection,\ and\ the\ other\ obtained\ by\ feedback\ interconnection,\ are\ both\ passive.\ If\ systems\ H_1\ and\ H_2\ are\ ZSD\ and\ their\ respective\ storage\ functions\ S_1(x_1)\ and\ S_2(x_2)\ are\ C^1,\ then\ the\ equilibrium\ (x_1, x_2) = (0, 0)\ of\ both\ interconnections\ is\ stable.$

$Proof.$ Passivity: Because H_1 and H_2 are passive, there exist two positive semidefinite storage functions $S_1(x_1)$ and $S_2(x_2)$ such that

$$S_i\left(x_i\left(t_1\right)\right) - S_i\left(x_i\left(t_0\right)\right) \leq \int_{t_0}^{t_1} u_i^T y_i dt, \quad i = 1, 2, \tag{2.66}$$

where x_1, x_2 are the state variables of H_1 and H_2, respectively. Define $x = \left[x_1^T, x_2^T\right]^T$ and $S\left(x\right) = S_1\left(x_1\right) + S_2\left(x_2\right)$. Note that $S\left(x\right)$ is positive semidefinite and

$$S\left(x\left(t_1\right)\right) - S\left(x\left(t_0\right)\right) \leq \int_{t_0}^{t_1} \left(u_1^T y_1 + u_2^T y_2\right) dt. \tag{2.67}$$

For the parallel interconnection, $u = u_1 = u_2$ and $y = y_1 + y_2$. Therefore,

$$S\left(x\left(t_1\right)\right) - S\left(x\left(t_0\right)\right) \leq \int_{t_0}^{t_1} u^T y dt. \tag{2.68}$$

For the feedback case, $u_2 = y_1$ and $u_1 = r - y_2$:

$$S\left(x\left(t_1\right)\right) - S\left(x\left(t_0\right)\right) \leq \int_{t_0}^{t_1} r^T y_1 dt. \tag{2.69}$$

Therefore, both interconnections are passive.

If systems H_1 and H_2 are ZSD, the equilibrium $(x_1, x_2) = (0,0)$ of both interconnections is Lyapunov stable, according to Theorem 2.11.

The above conditions can be extended to partial parallel and feedback connections:

Proposition 2.34 (Partial interconnection of passive systems). *Consider systems $H_1 : u_1 \longmapsto y_1$ and $H_2 : u_2 \longmapsto y_2$, where $u_1 = \left[u_{11}^T, u_{12}^T\right]^T$, $u_2 = \left[u_{21}^T, u_{22}^T\right]^T$, $y_1 = \left[y_{11}^T, y_{12}^T\right]^T$, $y_2 = \left[y_{21}^T, y_{22}^T\right]^T$. If systems H_1 and H_2 are passive, then the two systems, one obtained by partial parallel interconnection, and the other obtained by partial feedback interconnection (as shown in Figure 2.4), are both passive. If systems H_1 and H_2 are ZSD and their respective storage functions $S_1(x_1)$ and $S_2(x_2)$ are C^1, then the equilibrium $(x_1, x_2) = (0,0)$ of both interconnections is stable.*

Proof. Similar to the proof of Theorem 2.33, because H_1 and H_2 are passive, there exist two positive semidefinite storage functions $S_1\left(x_1\right)$ and $S_2\left(x_2\right)$ such that

$$S_i\left(x_i\left(t_1\right)\right) - S_i\left(x_i\left(t_0\right)\right) \leq \int_{t_0}^{t_1} u_i^T y_i dt, \quad i = 1, 2, \tag{2.70}$$

where x_1, x_2 are the state variables of H_1 and H_2, respectively. Define $x = \left[x_1^T, x_2^T\right]^T$ and $S\left(x\right) = S_1\left(x_1\right) + S_2\left(x_2\right)$. Note that $S\left(x\right)$ is positive semidefinite and

$$\begin{aligned}
S\left(x\left(t_1\right)\right) - S\left(x\left(t_0\right)\right) &\leq \int_{t_0}^{t_1} \left(u_1^T y_1 + u_2^T y_2\right) dt \\
&= \int_{t_0}^{t_1} \left(u_{11}^T y_{11} + u_{12}^T y_{12} + u_{21}^T y_{21} + u_{22}^T y_{22}\right) dt.
\end{aligned} \tag{2.71}$$

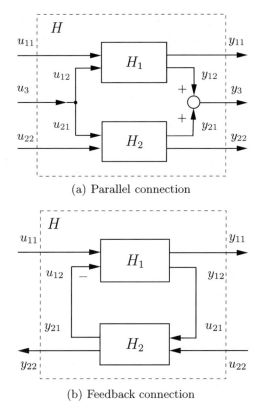

(a) Parallel connection

(b) Feedback connection

Fig. 2.4. Partial interconnected systems

For the partial parallel interconnection, define $u_3 = u_{12} = u_{21}$, $y_3 = y_{12} + y_{21}$. The overall system inputs and outputs are $u = \left[u_{11}^T, u_3^T, u_{22}^T\right]^T$ and $y = \left[y_{11}^T, y_3^T, y_{22}^T\right]^T$, respectively. Therefore,

$$
\begin{aligned}
S\left(x\left(t_1\right)\right) - S\left(x\left(t_0\right)\right) &\leq \int_{t_0}^{t_1} \left(u_{11}^T y_{11} + u_3^T y_3 + u_{22}^T y_{22}\right) dt \\
&= \int_{t_0}^{t_1} y^T u \, dt.
\end{aligned}
\tag{2.72}
$$

For the feedback case, $u_{12} = -y_{21}$ and $u_{21} = y_{12}$. The overall system inputs and outputs are $u = \left[u_{11}^T, u_{22}^T\right]^T$ and $y = \left[y_{11}^T, y_{22}^T\right]^T$, respectively. Then

$$
\begin{aligned}
S\left(x\left(t_1\right)\right) - S\left(x\left(t_0\right)\right) &\leq \int_{t_0}^{t_1} \left(u_{11}^T y_{11} - y_{21}^T y_{12} + y_{12}^T y_{21} + u_{22}^T y_{22}\right) dt \\
&= \int_{t_0}^{t_1} \left(u_{11}^T y_{11} + u_{22}^T y_{22}\right) dt = \int_{t_0}^{t_1} y^T u \, dt.
\end{aligned}
\tag{2.73}
$$

Therefore, both interconnections are passive. If systems H_1 and H_2 are ZSD, from Theorem 2.11, the equilibrium $(x_1, x_2) = (0, 0)$ of both interconnections is Lyapunov stable.

As a result, if a process is passive, it can be stabilized at the equilibrium point $(x = 0)$ by any passive controller, even if it is highly nonlinear and/or highly coupled. For example, the gravity tank can be stabilized by *any* PID controller with a positive controller gain. The controller gain can be arbitrarily large to reduce the response time without causing instability. This motivates stability analysis and control design based on passivity. The above stability condition can be further extended by introducing the notion of a passivity index.

2.4 Passivity Indices

2.4.1 Excess and Shortage of Passivity

To extend the passivity-based stability conditions to more general cases for both passive and nonpassive systems, we need to define the passivity indices that quantify the degree of passivity. The passivity indices can be defined in terms of an excess or shortage of passivity.

Let system H, as given in (2.32), be passive with a C^1 storage function $S(x)$. Consider a static feedfoward $y_{ff} = -\nu u$ $(\nu > 0)$ such that the overall system $\tilde{H}$ has the output $\tilde{y} = y - \nu u$ (as shown in Figure 2.5a). Because the feedforward is static, its state-space is void. Therefore, the storage function of the overall system remains $S(x)$. If $\tilde{H}$ is also passive, then,

$$S(x(t_1)) - S(x(t_0)) \le \int_{t_0}^{t_1} u^T \tilde{y} dt = \int_{t_0}^{t_1} \left(u^T y - \nu u^T u \right) dt. \qquad (2.74)$$

This is equivalent to the condition that H is dissipative with respect to the supply rate $w(u, y) = u^T y - \nu u^T u$. In this case, system H is said to have excessive input feedforward passivity of ν, denoted as IFP(ν). The feedforward system $-\nu I$ is not passive because $\int_{t_0}^{t_1} u^T y_{ff} dt = \int_{t_0}^{t_1} -\nu u^T u dt < 0$, violating the positive real condition. From this example, it can be seen that the excess of passivity in H can compensate for the shortage of passivity in the feedforward system. Similarly, if H is nonpassive, but it is dissipative with respect to the supply rate $w(u, y) = u^T y + \nu u^T u$ $(\nu > 0)$, then system $H + \nu I$ is passive. In this case, H lacks input feedforward passivity, denoted as IFP($-\nu$).

Another situation is the negative feedback interconnection (as shown in Figure 2.5b). Let $\tilde{H}$ be the closed-loop system of H with a positive feedback ρI $(\rho > 0)$. Assume that $\tilde{H}$ is passive with a C^1 storage function $S(x)$, then,

$$S(x(t_1)) - S(x(t_0)) \le \int_{t_0}^{t_1} r^T y dt = \int_{t_0}^{t_1} \left(u^T y - \rho y^T y \right) dt. \qquad (2.75)$$

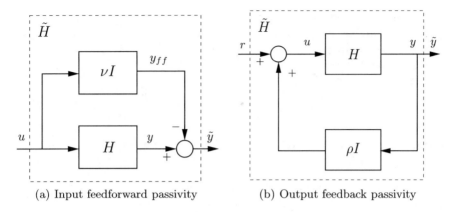

(a) Input feedforward passivity   (b) Output feedback passivity

Fig. 2.5. Excess and shortage of passivity

This is equivalent to the dissipativity of system H with respect to the supply rate $w(u, y) = u^T y - \rho y^T y$. In this case, H is said to have excessive output feedback passivity of ρ, denoted as OFP(ρ). If H is not passive, but it is dissipative with respect to the supply rate $w(u, y) = u^T y + \rho y^T y$ ($\rho > 0$), then system H can be rendered passive by a *negative* feedback ρI. In this case, H is said to lack output feedback passivity, denoted as OFP($-\rho$). Mathematically,

Definition 2.35 (Excess/shortage of passivity [110]). *Let $H : u \longmapsto y$. System H is said to be:*

1. *Input feedforward passive (IFP) if it is dissipative with respect to supply rate $w(u, y) = u^T y - \nu u^T u$ for some $\nu \in \mathbb{R}$, denoted as IFP(ν).*
2. *Output feedback passive (OFP) if it is dissipative with respect to supply rate $w(u, y) = u^T y - \rho y^T y$ for some $\rho \in \mathbb{R}$, denoted as OFP(ρ).*

In this book, a positive value of ν or ρ means that the system has an excess of passivity. In this case, the process is said to be *strictly input passive* or *strictly output passive*, respectively. Clearly, if a system is IFP(ν) or OFP(ρ), then it is also IFP($\nu - \varepsilon$), or OFP($\rho - \varepsilon$) $\forall \varepsilon > 0$.

The IFP and OFP can also be defined on the input-output version of passivity:

Definition 2.36 ([130]). *Let $H : \mathcal{L}_{2e}^m \to \mathcal{L}_{2e}^m$. System H is strictly input passive if there exist β and $\delta > 0$ such that*

$$\langle Hu, u \rangle_T \geq \delta \|u_T\|_2^2 + \beta, \ \forall \, u \in \mathcal{L}_{2e}^m, \ T \geq 0. \tag{2.76}$$

H is strictly output passive if there exist β and $\varepsilon > 0$ such that

$$\langle Hu, u \rangle_T \geq \varepsilon \|(Hu)_T\|_2^2 + \beta, \ \forall \, u \in \mathcal{L}_{2e}^m, \ T \geq 0. \tag{2.77}$$

A strictly output passive system has a finite $\mathcal{L}_2$ gain [130]. Furthermore, a system that has excessive OFP with a C^1 storage function has a stable equilibrium $x = 0$ when $u = 0$, provided that the system is ZSD. This can be seen from the following:

$$
\begin{aligned}
\dot{S} &\leq u^T y - \rho y^T y = u^T y - \rho h^T(x) h(x) \\
&< -\rho h^T(x) h(x) < 0, \quad \forall \, h(x) \neq 0 \text{ and } u = 0.
\end{aligned}
\tag{2.78}
$$

Following a proof similar to Theorem 2.32, $x \to 0$ when $t \to \infty$.

IFP and OFP systems have the following scaling property:

Proposition 2.37 (IFP/OFP Scaling [110]). *For systems H and αH, where α is a constant, the following statements are true:*

1. *If H is OFP(ρ), then αH is OFP($\frac{1}{\alpha}\rho$).*
2. *If H is IFP(ν), then αH is IFP($\alpha\nu$).*

Note that the strict passivity definition for linear systems given in Theorem 2.25 is the IFP plus the stability condition, not the linear version of state strict passivity for nonlinear systems. More precisely, a linear system is *strictly passive* if it is *stable* and IFP(ν), $\nu > 0$.

Example 2.38. To illustrate the definition of IFP and OFP, let us consider a linear integrating system:

$$
H : \begin{cases} \dot{x} &= u \\ y &= x. \end{cases}
\tag{2.79}
$$

This system is lossless (passive but not strictly passive). By definition, system H with a positive feedforward ν:

$$
H_1 : \begin{cases} \dot{x} &= u \\ y &= x + \nu u \end{cases}
\tag{2.80}
$$

will have excessive IFP of ν. This can be seen by using a storage function $S(x) = \frac{1}{2}x^2$:

$$
\dot{S} = xu = yu - \nu u^2.
\tag{2.81}
$$

From an input-output point of view, $H(s) = 1/s$ is passive, and

$$
H_1(s) = H(s) + \nu = (\nu s + 1)/s
\tag{2.82}
$$

has excessive IFP of ν. According to Theorem 2.25, $H_1(s)$ is not strictly passive because it is not stable.

Similarly, $H(s)$ with a negative feedback of ρ ($\rho > 0$),

$$
H_2(s) = \frac{\frac{1}{s}}{1 + \rho \frac{1}{s}} = \frac{1}{s + \rho},
\tag{2.83}
$$

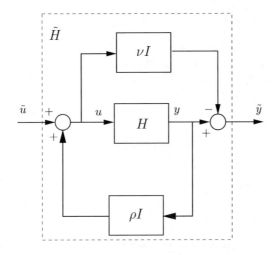

Fig. 2.6. Simultaneous IFP and OFP

will be OFP(ρ) and also strictly passive. Linear strictly output passive systems may not be strictly passive (due to the fact that strict passivity for linear systems requires strict IFP). For example,

$$H_3(s) = \frac{s}{s+1} \tag{2.84}$$

is OFP(1), but it is not strictly passive because $H_3(0) + H_3^*(0) = 0$.

More general supply rates can be used to define simultaneous IFP and OFP. Consider a system H with both input feedforward νI and output feedback ρI, as shown in Figure 2.6. If the overall system $\tilde{H}$ is passive, then system H is dissipative with respect to the supply rate:

$$w(u,y) = (1 + \rho\nu) y^T u - \nu u^T u - \rho y^T y. \tag{2.85}$$

In the above discussion, the feedforward and feedback are assumed to be static and decentralized. A more general case is when they are arbitrary nonlinear multivariable (thus vector) functions, *e.g.*

$$w(u,y) = y^T u - \nu^T(u) u - \rho^T(y) y, \tag{2.86}$$

where $v(u) = [v_1(u), \cdots v_m, (u)]^T$ and $\rho(u) = [\rho_1(u), \cdots, \rho_m(u)]^T$.

Another generalization of the supply rate was given by Hill and Moylan [57]:

$$w(u(t), y(t)) = y^T(t) Q y(t) + 2u^T(t) S y(t) + u^T(t) R u(t), \tag{2.87}$$

where $Q, R, S \in \mathbb{R}^{m \times m}$ are constant weighting matrices, with Q and R symmetrical. This corresponds to multivariable but linear and static feedforward and feedback required to render the process system passive.

2.4.2 Passivity Indices for Linear Systems

For a stable linear system with a transfer function $G(s)$, the IFP index, denoted as $\nu(G(s))$, can be calculated based on the KYP lemma. If $G(s)$ has excessive IFP, then there exists a largest $\nu > 0$ such that the process with the feedforward $-\nu I$ is positive real, *i.e.*,

$$G(j\omega) - \nu I + [G(j\omega) - \nu I]^* > 0, \quad \forall \, \omega. \tag{2.88}$$

Therefore, we can have the following definition:

Definition 2.39. *The input feedforward passivity index for a stable linear system $G(s)$ is defined as*[2]

$$\nu(G(s)) \triangleq \frac{1}{2} \min_{\omega \in \mathbb{R}} \underline{\lambda}(G(j\omega) + G^*(j\omega)), \tag{2.89}$$

where $\underline{\lambda}$ denotes the minimum eigenvalue.

If ν is negative, then the minimum feedforward required to render the process passive is νI. The above definition also gives a numerical approach for calculating the IFP index. For linear systems, it is possible to define a tighter IFP index conveniently by employing a frequency-dependent passivity index:

Definition 2.40 ([11]). *The input feedforward passivity index for a stable linear system $G(s)$ at frequency ω is given by*

$$\nu_F(G(s), \omega) \triangleq \frac{1}{2} \underline{\lambda}(G(j\omega) + G^*(j\omega)). \tag{2.90}$$

By using the above definition, we can specify the condition that a dynamic feedforward $G_{ff}(s)$ needs to satisfy so that $G(s) + G_{ff}(s)$ is passive. For a stable process $G(s)$, a stable $G_{ff}(s)$ should be chosen such that

$$\nu_F(G_{ff}(s), \omega) + \nu_F(G(s), \omega) > 0 \quad \forall \, \omega \in \mathbb{R}. \tag{2.91}$$

It is more difficult to calculate the OFP index numerically because it involves feedback loops. If a process $G(s)$ is minimum phase (therefore, $G^{-1}(s)$ exists and is stable. $G(s)$ does not need to be stable), then with a positive feedback of ρI, the closed-loop system is

$$G_{cl}(s) = G(s)[I - \rho G(s)]^{-1} = [G(s)^{-1} - \rho I]^{-1}. \tag{2.92}$$

According to Proposition 2.23, $G_{cl}(s)$ is passive if and only if

$$G_{cl}^{-1}(s) = G(s)^{-1} - \rho I \tag{2.93}$$

is passive. Therefore, the OFP index of $G(s)$ is the IFP index of $G^{-1}(s)$. We can have the following definition:

---

[2] This definition is similar to the passivity index proposed in [135], except that in [135] a positive value of ν implies that the system lacks passivity.

Definition 2.41. *The output feedback passivity index for a minimum phase linear system $G(s)$ is defined as*

$$\rho\left(G\left(s\right)\right) \triangleq \frac{1}{2} \min_{\omega \in \mathbb{R}} \lambda\left(G^{-1}\left(j\omega\right) + \left[G^{-1}\left(j\omega\right)\right]^{*}\right). \tag{2.94}$$

The OFP index at frequency ω is given by

$$\rho_{F}\left(G\left(s\right), \omega\right) \triangleq \frac{1}{2} \lambda\left(G^{-1}\left(j\omega\right) + \left[G^{-1}\left(j\omega\right)\right]^{*}\right). \tag{2.95}$$

For processes that are nonminimum phase and unstable, we need both feedback and feedforward to render the process passive. In this case, the IFP and OFP indices are dependent. Special passivity indices need to be defined so that they can be conveniently computed and used in system analysis and control design. We will introduce these indices in other chapters.

2.5 Passivation

To render a process passive via either feedback or feedforward is called *passivation*. This is possible if the process lacks either IFP or OFP. Because passive systems are stable and easy to control, passivation is often a useful step in control design. For example, we may passivate a process and then stabilize the passivated system with a (strictly) passive controller (*e.g.*, a static output feedback controller given in Theorem 2.32).

2.5.1 Input Feedforward Passivation

Many stable processes can be passivated by a static feedforward. For example, a linear system $G_1\left(s\right) = \frac{1-s}{s^3+s^2+s+1}$ can be passivated by a static unit feedforward because $G\left(s\right) = G_1\left(s\right) + 1 = \frac{s^3+s^2+2}{s^3+s^2+s+1}$ is minimum phase, has a relative degree of 0 and is positive real.

Consider a control affine process H as in (2.32). Assume that the process has a globally stable equilibrium at $x = 0$ with a Lyapunov function $V\left(x\right)$. Use $V\left(x\right)$ as a storage function, then

$$\frac{\mathrm{d}V\left(x\right)}{\mathrm{d}t} = \frac{\partial V\left(x\right)}{\partial x} f\left(x\right) + \frac{\partial V\left(x\right)}{\partial x} g\left(x\right) u \leq \frac{\partial V\left(x\right)}{\partial x} g\left(x\right) u. \tag{2.96}$$

As shown in Figure 2.5a with the feedforward νI, $\tilde{y} = h\left(x\right) + \nu u$. Then $\tilde{y}^T u = h^T\left(x\right) u + \nu u^T u$. As long as there exists a ν such that

$$\nu u^T u > \left[\frac{\partial V\left(x\right)}{\partial x} g\left(x\right) - h^T\left(x\right)\right] u, \tag{2.97}$$

$\dot{V}\left(x\right) \leq \tilde{y}^T u.$

This result can be generalized to dynamic feedforward systems. Any stable control affine process (of which a Lyapunov function can be found) can be passivated with a feedforward dynamic system. As shown in Figure 2.3a, assume that a system

$$H_1 : \begin{cases} \dot{x} &= f_1(x) + g_1(x) u_1 \\ y_1 &= h_1(x), \end{cases} \tag{2.98}$$

is nonpassive but has a globally stable equilibrium point $x = 0$ with a Lyapunov function $V(x)$. A feedforward system H_2 can be designed to passivate H_1. One way to design such a feedfoward passivater is to assume that the passivated system H has the same state equation as that of H_1 and find an appropriate output function $y(t) = h(x)$ such that H is passive. According to the KYP lemma (Proposition 2.14), if we use $V(x)$ as a storage function, then, the condition $L_f V(x) = \frac{\partial V(x)}{\partial x} f_1(x) \leq 0$ is always satisfied. If we choose $h(x) = \left[\frac{\partial V(x)}{\partial x} g_1(x) \right]^T$, then H is passive. The feedforward system H_2 can be obtained by subtracting y from y_1:

$$H_2 : \begin{cases} \dot{x} &= f_1(x) + g_1(x) u_2 \\ y_2 &= \left[\frac{\partial V(x)}{\partial x} g_1(x) \right]^T - h_1(x). \end{cases} \tag{2.99}$$

Such a feedforward will stabilize the zero dynamics of H_1 (so that H is made weakly minimum phase) and reduce its relative degree to no greater than $\{1, \cdots, 1\}$.

For linear systems, the feedforward system can be easily obtained using the linear version of the KYP lemma. Detailed discussion will be given in later chapters. However, it is not possible to passivate an unstable process with feedforward because the feedforward does not affect the free dynamics of the process (when $u = 0$). Such systems can only be passivated via feedback.

2.5.2 Output Feedback Passivation

Passivation of unstable processes is a topic which attracted much interest because it can be an effective approach to stabilization of nonlinear processes. Most research work is concerned with passivation by state feedback. A thorough development of this topic can be found in [24]. A control affine system given in (2.32) is said to be *feedback passive* (or *feedback equivalent to a passive system*) if there exists a state feedback transformation [24]:

$$u = \alpha(x) + \beta(x) v, \tag{2.100}$$

with invertible $\beta(x)$ such that the system

$$\dot{x} = f(x) + g(x) \alpha(x) + g(x) \beta(x) v, \\ y = h(x), \tag{2.101}$$

is passive. The condition for feedback passivity is given in the following theorem:

Theorem 2.42 (State feedback passivity [24]). *Consider the control affine system in (2.32). Assume* rank $\left(L_g h\left(x\right)|_{x=0}\right) = m$ *(where m is the number of outputs). Then this system is feedback passive with a C^2 positive definite storage function $S\left(x\right)$ if and only of it has relative degree $\{1,\cdots 1,\}$ at $x = 0$ and is weakly minimum phase.*

Clearly, the above condition says that we cannot render a nonminimum phase system or a system with a relative degree larger than 1 passive via feedback, because a passive system needs to be weakly minimum phase and have a relative degree no greater than 1, but the relative degree and the zero dynamics *cannot* be altered by feedback [69]. In this case, passivation is only possible via feedforward.

For the output feedback case, an additional condition is required:

Theorem 2.43 (Output feedback passivity).

1. *Necessary condition: If the system in (2.32) can be rendered passive with a C^2 storage function $S\left(x\right)$, then it has relative degree $\{1,\cdots,1\}$ at $x = 0$ and is weakly minimum phase, and $L_g h\left(x\right)|_{x=0}$ is symmetrical and positive definite.*
2. *Sufficient condition: The system in (2.32) can be rendered locally passive with a C^2 positive definite storage function $S\left(x\right)$ by an output feedback if its Jacobian linearization at $x = 0$ is minimum phase and $\frac{\partial h(x)}{\partial x}g\left(x\right)\Big|_{x=0}$ is symmetrical and positive definite.*

To get some intuition from the above conditions, let us look at the case of linear systems. For a linear system

$$\dot{x} = Ax + Bu,$$
$$y = Cx, \tag{2.102}$$

$$L_g h\left(x\right) = \frac{\partial h\left(x\right)}{\partial x}g\left(x\right) = CB. \tag{2.103}$$

Theorem 2.42 says the linear system is feedback passive if it (1) has a relative degree of 1 (due to the assumption $D = 0$, the relative degree cannot be 0) (2) is weakly minimum phase (it may have zeros in the LHP and on the imaginary axis) and (3) rank $\left(CB\right) = m$. Note that if CB is nonsingular, then the linear system has a relative degree of 1. Therefore, Condition (3) implies Condition (1) for linear systems.

Clearly, any *state feedback* cannot change any of the above conditions, because with a state feedback, $u = r - Kx$ (r is an exogenous input such as reference), the closed-loop system will be

$$\dot{x} = (A - BK)x + Br,$$
$$y = Cx. \tag{2.104}$$

For example, systems like $G_1(s) = \frac{1-s}{s+1}$ cannot be passivated by any state feedback controllers.

For *output feedback* passivity, CB must also be (1) symmetrical and (2) positive definite. The first condition implies that the input and output of a process need to be properly paired. The second condition imposes the limitation on the sign of the steady-state gain. For example, a linear system $G_2(s) = \frac{-1}{-s+1}$ can be stabilized by a negative feedback controller $K(s)$ with negative steady-state gain, but cannot be made positive real because the closed-loop system $\frac{G_2(s)}{1+G_2(s)K(s)}$ will have a negative steady-state gain.

2.6 Passivity Theorem

We have shown the stability condition of passive systems in feedback in Theorem 2.33. By using the notions of strict input passivity and strict output passivity, asymptotic stability conditions for interconnected passive systems can be derived. These conditions are called the Passivity Theorem. The simplest version of the Passivity Theorem is as follows:

Theorem 2.44 (Passivity Theorem [110]). *Assume that systems H_1 and H_2 are ZSD and dissipative with C^1 storage functions $S_1(x_1)$ and $S_2(x_2)$. Then the equilibrium $(x_1, x_2) = (0,0)$ of their feedback connection (as shown in Figure 2.7a) with $r \equiv 0$ is asymptotically stable (AS) if*

1. *H_1 and H_2 are strictly output passive; or,*
2. *H_1 and H_2 are strictly input passive; or,*
3. *H_1 is GAS and strictly input passive and H_2 is passive.*

If storage functions $S_1(x_1)$ and $S_2(x_2)$ are radially unbounded, then the feedback connection is globally asymptotically stable (GAS).

Proof. The proof of the above theorem can be found in [110]. Here we provide a simplified version of the proof to clarify the intent. The storage function for the closed-loop is chosen as $S(x_1, x_2) = S_1(x_1) + S_2(x_2)$.

1. Since H_1 and H_2 are strictly output passive, there exist $\rho_1, \rho_2 > 0$ such that

$$\dot{S}_1(x_1) \leq y_1^T u_1 - \rho_1 y_1^T y_1, \tag{2.105}$$
$$\dot{S}_2(x_2) \leq y_2^T u_2 - \rho_2 y_2^T y_2. \tag{2.106}$$

Then,

$$\dot{S}(x_1, x_2) \leq y_1^T u_1 - \rho_1 y_1^T y_1 + y_2^T u_2 - \rho_2 y_2^T y_2. \tag{2.107}$$

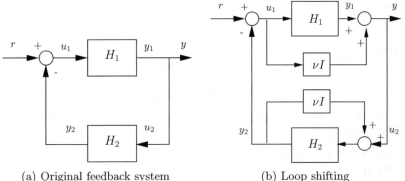

(a) Original feedback system   (b) Loop shifting

Fig. 2.7. Extended passivity condition

Because $u_1 = -y_2, u_2 = y_1$,

$$\dot{S}(x_1, x_2) \leq -y_1^T y_2 - \rho_1 y_1^T y_1 + y_2^T y_1 - \rho_2 y_2^T y_2$$
$$= -\rho_1 y_1^T y_1 - \rho_2 y_2^T y_2 < 0, \quad \forall \, y_1, y_2 \neq 0. \tag{2.108}$$

The bounded solution of (x_1, x_2) is confined in $\{(x_1, x_2) | (y_1, y_2) = (0, 0)\}$. Because H_1 and H_2 are ZSD, $(x_1, x_2) \to (0, 0)$.

2. Since H_1 and H_2 are strictly input passive, there exist $\nu_1, \nu_2 > 0$ such that

$$\dot{S}_1(x_1) \leq y_1^T u_1 - \nu_1 u_1^T u_1, \tag{2.109}$$

$$\dot{S}_2(x_2) \leq y_2^T u_2 - \nu_2 u_2^T u_2. \tag{2.110}$$

Then

$$\dot{S}(x_1, x_2) \leq y_1^T u_1 - \nu_1 u_1^T u_1 + y_2^T u_2 - \nu_2 u_2^T u_2$$
$$\leq -\nu_1 y_2^T y_2 - \nu_2 y_1^T y_1 < 0, \quad \forall \, y_1, y_2 \neq 0. \tag{2.111}$$

Similar to Part 1, $(x_1, x_2) \to (0, 0)$.

3. In this case, there exists a $\nu_1 > 0$ such that

$$\dot{S}_1(x_1) \leq y_1^T u_1 - \nu_1 u_1^T u_1, \tag{2.112}$$

$$\dot{S}_2(x_2) \leq y_2^T u_2, \tag{2.113}$$

$$\dot{S}(x_1, x_2) \leq y_1^T u_1 - \nu_1 u_1^T u_1 + y_2^T u_2$$
$$= -\nu_1 y_2^T y_2 < 0, \quad \forall \, y_2 \neq 0. \tag{2.114}$$

Because $\dot{S}$ is bounded only by $y_2^T y_2$, the bounded solution of (x_1, x_2) is confined in $\{(x_1, x_2) | y_2 = 0\}$ and $u_1 = 0$. Because H_1 is GAS and H_2 is ZSD, $(x_1, x_2) \to (0, 0)$.

If storage functions $S_1(x_1)$ and $S_2(x_2)$ are radially unbounded, then *all* the above results hold globally.

The input-output version of the Passivity Theorem can be presented as follows:

Theorem 2.45 ([130]). *Consider the closed-loop system shown in Figure 2.7a with H_1, $H_2 : \mathcal{L}_{2e}^m \to \mathcal{L}_{2e}^m$. Assume that for any $r \in \mathcal{L}_2^m$ there are solutions u_1, $u_2 \in \mathcal{L}_{2e}^m$. If*

1. *H_1 is passive and H_2 is strictly input passive; or,*
2. *H_1 is strictly output passive and H_2 is passive,*

then, $u_2 = y_1 = H_1(u_1) \in \mathcal{L}_2^m$, i.e., the closed-loop system from r to y_1 is $\mathcal{L}_2$ stable.

Furthermore, if the input-output stability of systems H_1 and/or H_2 is assumed, we have

Theorem 2.46 ([130]). *Consider the closed-loop system shown in Figure 2.7a with H_1, $H_2 : \mathcal{L}_{2e}^m \to \mathcal{L}_{2e}^m$. Assume for any $r \in \mathcal{L}_2^m$ that there are solutions u_1, $u_2 \in \mathcal{L}_{2e}^m$. If*

1. *H_1 is passive and H_2 is strictly input passive and $\mathcal{L}_2$ stable; or*
2. *Both H_1 and H_2 are strictly output passive,*

then, $y_1, y_2 \in \mathcal{L}_2^m$, i.e., both of the closed-loop systems from r to y_1 and from r to y_2 are $\mathcal{L}_2$ stable.

For linear systems, Condition 1 of the above theorem simply means:

Proposition 2.47 (Passivity theorem for linear systems). *Consider two LTI systems H_1 and H_2 in negative feedback configuration, as shown in Figure 2.7a. The closed-loop system is asymptotically stable if H_1 is strictly passive and H_2 is passive.*

This can be clearly seen from the example of two SISO systems H_1 and H_2. In this case, the phase shifts of H_1 and H_2 lie within $(-90°, 90°)$ and $[-90°, 90°]$, respectively. Therefore, the total phase shift of the open-loop system never reaches $-180°$, producing no critical frequency in the open-loop Bode diagram. According to the Nyquist-Bode stability condition, the closed-loop system is stable *regardless* of the amplitude ratio of $H_1(j\omega)H_2(j\omega)$. The system has infinite gain margin.

By using the concepts of excess and shortage of passivity, we can extend the above results further to general (possibly nonpassive) systems. Assume that system H_1 in Figure 2.7a is GAS but lacks IFP, *e.g.*, is IFP$(-\nu_1)$, $\nu_1 > 0$, then a feedforward of νI (where $\nu = \nu_1 + \varepsilon$ and ε is an arbitrarily small positive number) will render H_1 strictly input passive, as depicted in Figure 2.7b. To make the feedback system equivalent to the original system in Figure 2.7a, a positive feedback of νI is added to H_2. According to Theorem 2.44, the equilibrium $(x_1, x_2) = (0,0)$ of the closed-loop system is GAS if H_2 with positive feedback is passive, i.e., H_2 has excessive output feedback passivity of ν. Similarly, a shortage of output feedback passivity of H_2 can be compensated for by excessive input feedforward passivity of H_1 so that the closed-loop system is GAS. More rigorously, we have:

Theorem 2.48 ([110]). *Assume that in the feedback interconnection shown in Figure 2.7a, H_1 is GAS and IFP(ν) and the system H_2 is ZSD and OFP(ρ). Then $(x_1, x_2) = (0,0)$ is AS if $\nu + \rho > 0$. If, in addition, the storage functions of H_1 and H_2 are radially unbounded, then $(x_1, x_2) = (0,0)$ is GAS.*

If the systems are characterized by a more general supply rate as in (2.86), the above condition can be further extended:

Theorem 2.49 ([110]). *Assume that the systems H_1 and H_2 are dissipative with respect to the following supply rates:*

$$w_i(u_i, y_i) = u_i^T y_i - \rho_i^T(y_i) y_i - \nu_i^T(u_i)(u_i), \quad i = 1, 2, \tag{2.115}$$

where $u_i, y_i \in \mathbb{R}^m$, $i = 1, 2$. Furthermore assume that they are ZSD and that their respective storage functions $S_1(x_1)$ and $S_2(x_2)$ are C^1. Then the equilibrium $(x_1, x_2) = (0,0)$ of the feedback interconnection in Figure 2.7a is

1. *stable, if $\nu_1^T(v)v + \rho_2^T(v)v \geq 0$ and $\nu_2^T(v)v + \rho_1^T(v)v \geq 0$, $\forall\, v \in \mathbb{R}^m$;*
2. *asymptotically stable, if $\nu_1^T(v)v + \rho_2^T(v)v > 0$ and $\nu_2^T(v)v + \rho_1^T(v)v > 0$, $\forall\, v \in \mathbb{R}^m$ and $v \neq 0$.*

One special case of the supply rates is $\nu_i(u_i) = \bar{\nu}_i u_i$ and $\rho_i(y_i) = \bar{\rho}_i y_i$, where $\bar{\nu}_i$ and $\bar{\rho}_i$ are scalar constants. In this case,

$$\nu_1^T(v)v + \rho_2^T(v)v = \bar{\nu}_1 v^T v + \bar{\rho}_2 v^T v = (\bar{\nu}_1 + \bar{\rho}_2)\, v^T v, \tag{2.116}$$

$$\nu_2^T(v)v + \rho_1^T(v)v = \bar{\nu}_2 v^T v + \bar{\rho}_1 v^T v = (\bar{\nu}_2 + \bar{\rho}_1)\, v^T v. \tag{2.117}$$

Then, the equilibrium $(x_1, x_2) = (0,0)$ of the feedback interconnection is

1. stable if $\bar{\nu}_1 + \bar{\rho}_2 \geq 0$ and $\bar{\nu}_2 + \bar{\rho}_1 \geq 0$;
2. asymptotically stable if $\bar{\nu}_1 + \bar{\rho}_2 > 0$ and $\bar{\nu}_2 + \bar{\rho}_1 > 0$.

Another special case is

$$\nu_1(v) = \rho_1(v) = 0, \quad v_2(v) = \nu v \text{ and } \rho_2(v) = \rho v. \tag{2.118}$$

This leads to the following stability condition:

Proposition 2.50. *Assume that H_1 is passive (i.e., dissipative with respect to the supply rate $w_1 = u_1^T y_1$) and H_2 is dissipative with respect to the supply rate of $w_2 = u_2^T y_2 - \rho y_2^T y_2 - \nu u_2^T u_2$. Assume that systems H_1 and H_2 are ZSD and their respective storage functions $S_1(x_1)$ and $S_2(x_2)$ are C^1. Then, the equilibrium $(x_1, x_2) = (0,0)$ of the feedback interconnection in Figure 2.7a is asymptotically stable if $\rho > 0$ and $\nu > 0$.*

This condition does not require system H_1 to be AS.

2.7 Heat Exchanger Example

Some process systems are inherently passive (after proper rescaling of the inputs and/or outputs). One of the examples is the heat exchanger, a device built for efficient heat transfer from one fluid to another. The fluids are separated by a solid wall so that they never mix. Heat exchangers are widely used in air conditioning, refrigeration, space heating, power production, and in virtually every chemical plant.

Consider a single tube-in-shell heat exchanger as depicted in Figure 2.8, where cooling water is used to remove heat from a process stream. The volumetric flow rates of the process (hot) and service (cold) streams are v_h and v_c. The inlet and outlet temperatures of the hot and cold streams are T_{hi}, T_{ho}, T_{ci} and T_{co}, respectively. Strictly speaking, a tube-in-shell heat exchanger is a distributed parameter system (which can be represented by partial differential equations), because the temperatures of the hot and cold streams in the tube are functions of the location in the tube. To simplify our discussion, an approximate lumped parameter model given by Hangos *et al.* [54] is adopted. The model was built under the following assumptions:

1. Constant volume of the hot and cold streams in the heat exchanger (V_h and V_c);
2. Constant physicochemical properties, including density of the hot and cold streams (ρ_h and ρ_c) and their specific heat (c_{Ph} and c_{Pc});
3. Constant heat transfer coefficient U and area A;
4. Both hot and cold streams are well mixed and the temperatures of the hot and cold streams inside the tube are approximated by the outlet temperatures T_{ho} and T_{co}.

The state equations of the heat exchanger can be developed based on energy balance [54]:

$$\dot{T}_{co}(t) = \frac{v_c(t)}{V_c}\left[T_{ci}(t) - T_{co}(t)\right] + \frac{UA}{c_{Pc}\rho_c V_c}\left[T_{ho}(t) - T_{co}(t)\right], \qquad (2.119)$$

$$\dot{T}_{ho}(t) = \frac{v_h(t)}{V_h}\left[T_{hi}(t) - T_{ho}(t)\right] + \frac{UA}{c_{Ph}\rho_h V_h}\left[T_{co}(t) - T_{ho}(t)\right]. \qquad (2.120)$$

The inputs of the above process are the inlet temperatures and flow rates of the hot and cold streams. The outputs and states are the outlet temperatures. Depending on the choices of the manipulated variables, different models can be derived.

Example 2.51 (Linear model). If the inlet temperatures are manipulated to control the outlet temperatures, with the assumption that the flow rates of the cold and hot streams are constant, a linear model can be derived:

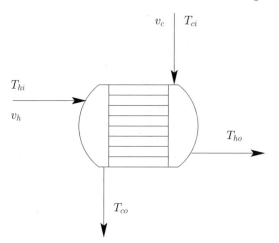

Fig. 2.8. A heat exchanger

$$\dot{x}(t) = \begin{bmatrix} -\frac{v_c}{V_c} - \frac{UA}{c_{Pc}\rho_c V_c} & \frac{UA}{c_{Pc}\rho_c V_c} \\ \frac{UA}{c_{Ph}\rho_h V_h} & -\frac{v_h}{V_h} - \frac{UA}{c_{Ph}\rho_h V_h} \end{bmatrix} x(t) + \begin{bmatrix} \frac{v_c}{V_c} & 0 \\ 0 & \frac{v_h}{V_h} \end{bmatrix} u(t), \quad (2.121)$$

$$y(t) = x(t), \quad (2.122)$$

where $x = [x_1, x_2]^T = [T_{co}, T_{ho}]^T$ and $u = [u_1, u_2]^T = [T_{ci}, T_{hi}]^T$. Define the following constants $k_1 = \frac{UA}{c_{Pc}\rho_c V_c}$, $k_2 = \frac{UA}{c_{Ph}\rho_h V_h}$, $a_1 = \frac{v_c}{V_c}$ and $a_2 = \frac{v_h}{V_h}$. Clearly, these constants are positive for any design and operating conditions. Then, the state equation becomes

$$\dot{x} = \begin{bmatrix} -a_1 - k_1 & k_1 \\ k_2 & -a_2 - k_2 \end{bmatrix} x + \begin{bmatrix} a_1 & 0 \\ 0 & a_2 \end{bmatrix} u. \quad (2.123)$$

To study the passivity of the above system, we define the following storage function:

$$S(x) = \frac{1}{2} x^T \begin{bmatrix} \frac{1}{k_1} & 0 \\ 0 & \frac{1}{k_2} \end{bmatrix} x > 0, \quad \forall\, x \neq 0. \quad (2.124)$$

Therefore,

$$\begin{aligned}
\dot{S}(x) &= x^T \left\{ \begin{bmatrix} \frac{1}{k_1} & 0 \\ 0 & \frac{1}{k_2} \end{bmatrix} \begin{bmatrix} -a_1 - k_1 & k_1 \\ k_2 & -a_2 - k_2 \end{bmatrix} x + \begin{bmatrix} a_1 & 0 \\ 0 & a_2 \end{bmatrix} u \right\} \\
&= x^T \begin{bmatrix} -\frac{a_1}{k_1} - 1 & 1 \\ 1 & -\frac{a_2}{k_2} - 1 \end{bmatrix} x + x^T \begin{bmatrix} \frac{a_1}{k_1} & 0 \\ 0 & \frac{a_2}{k_2} \end{bmatrix} u \\
&= -\frac{a_1}{k_1} x_1^2 - \frac{a_2}{k_2} x_2^2 - (x_1 - x_2)^2 + \frac{a_1}{k_1} x_1 u_1 + \frac{a_2}{k_2} x_2 u_2 \\
&\leq \frac{a_1}{k_1} y_1 u_1 + \frac{a_2}{k_2} y_2 u_2. \quad (2.125)
\end{aligned}$$

Note that the coefficients $k_1, k_2, a_1, a_2 > 0$. If the outputs are rescaled as $y^* = [y_1^*, y_2^*]^T = \left[\frac{a_1}{k_1} y_1, \frac{a_2}{k_2} y_2\right]^T$,

$$\dot{S}(x) < u^T y^*, \quad \forall\, x \neq 0, \tag{2.126}$$

leading to the conclusion that the heat exchanger is passive, regardless of design parameters (such as U, V_c, V_h, A), types of fluid (such as c_{Pc} and ρ_c) and operating conditions (such as v_c and v_h). If the heat exchanger parameters given in [65] are adopted, then $v_c = 2.29 \times 10^3$ ft^3/h, $v_h = 6.24 \times 10^3$ ft^3/h, $V_c = 5.57$ ft^3, $V_h = 20.40$ ft^3, $A = 521.5$ ft^2, $c_{Ph} = 0.58$ Btu/(lb·F), $c_{Pc} = 0.56$ Btu/(lb·F), $U = 75$ Btu/(h· ft^2·F), $\rho_h = 47.74$ lb/ft^3 and $\rho_c = 44.93$ lb/ft^3. In this case, (2.121) and (2.122) become

$$\dot{x} = \begin{bmatrix} -690.87 & 279.17 \\ 69.254 & -375.29 \end{bmatrix} x + \begin{bmatrix} 411.7 & 0 \\ 0 & 306.03 \end{bmatrix} u,$$

$$y = \begin{bmatrix} 1 & 0 \\ 0 & 1 \end{bmatrix} x. \tag{2.127}$$

It is easy to verify that the above process is passive, because matrices

$$P = \begin{bmatrix} 0.0024 & 0 \\ 0 & 0.0030 \end{bmatrix} \text{ and } L = \begin{bmatrix} 3.3562 & -0.9044 \\ -0.9044 & 2.4526 \end{bmatrix} \tag{2.128}$$

$$Q = W = 0 \tag{2.129}$$

are found to satisfy the conditions given in (2.41). The IFP index plot of this process is shown in Figure 2.9a. Its phase plot is given in Figure 2.9b, from which it can be seen that the phase shift is within $(-90°, 90°)$ at all frequencies.

Example 2.52 (Nonlinear model). A more realistic choice of manipulated variables is the flow rates of hot and cold streams, *i.e.*, $u = [u_1, u_2]^T = [v_h, v_c]^T$. In this case, we assume that the inlet temperatures T_{ci} and T_{hi} are constant. This leads to a nonlinear model. To study the passivity of the process with respect to an equilibrium point $x_0 = [x_{10}, x_{20}]^T = [T_{co0}, T_{hi0}]^T$, we define the following deviation variables: $x' = [x_1', x_2']^T = x - x_0$ and $u' = [u_1', u_2']^T = u - u_0$, where $u_0 = [v_{h0}, v_{c0}]^T$. (Note: The deviation variables can have negative values.) Therefore,

$$\dot{x}_1' = -k_1 (x_1' + x_{10}) + k_1 (x_2' + x_{20}) + \left[\frac{T_{ci}}{V_c} - \frac{1}{V_c}(x_1' + x_{10})\right](u_1' + u_{10}),$$

$$\dot{x}_2' = k_2 (x_1' + x_{10}) - k_2 (x_2' + x_{20}) + \left[\frac{T_{hi}}{V_h} - \frac{1}{V_h}(x_2' + x_{20})\right](u_2' + u_{20}). \tag{2.130}$$

Assume that (x_0, u_0) is at steady state, *i.e.*,

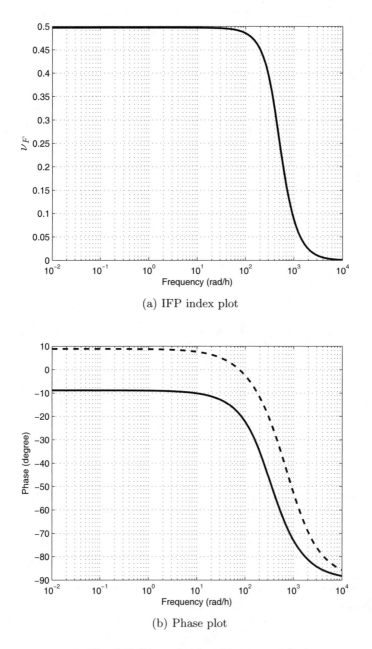

(a) IFP index plot

(b) Phase plot

Fig. 2.9. Linear heat exchanger model

$$0 = -k_1 x_{10} + k_1 x_{20} + \frac{T_{ci}}{V_c} u_{10} - \frac{1}{V_c} x_{10} u_{10},$$

$$0 = k_2 x_{10} - k_2 x_{20} + \frac{T_{hi}}{V_h} u_{20} - \frac{1}{V_h} x_{20} u_{20}. \tag{2.131}$$

Therefore,

$$\dot{x}_1' = -k_1 x_1' + k_1 x_2' + \frac{T_{ci}}{V_c} u_1' - \frac{1}{V_c} x_1' u_1' - \frac{1}{V_c} x_{10} u_1' - \frac{1}{V_c} x_1' u_{10},$$

$$\dot{x}_2' = k_2 x_1' - k_2 x_2' + \frac{T_{hi}}{V_h} u_2' - \frac{1}{V_h} x_2' u_2' - \frac{1}{V_h} x_{20} u_2' - \frac{1}{V_h} x_2' u_{20}. \tag{2.132}$$

Define a storage function

$$S(x') = \frac{1}{2} x'^T \begin{bmatrix} \frac{1}{k_1} & 0 \\ 0 & \frac{1}{k_2} \end{bmatrix} x' > 0, \quad \forall\, x' \neq 0; \tag{2.133}$$

then,

$$\dot{S}(x') = -(x_1' - x_2')^2 - \frac{1}{V_c k_1}(u_1' + u_{10}) x_1'^2 - \frac{1}{V_h k_2}(u_2' + u_{20}) x_2'^2$$

$$+ \left(\frac{T_{ci} - x_{10}}{V_c k_1} \right) x_1' u_1' + \left(\frac{T_{hi} - x_{20}}{V_h k_2} \right) x_2' u_2'. \tag{2.134}$$

Define a rescaled output $y^* = [y_1^*, y_2^*]^T = \left[\left(\frac{T_{ci} - x_{10}}{V_c k_1} \right) x_1', \left(\frac{T_{hi} - x_{20}}{V_h k_2} \right) x_2' \right]^T$. Also note $u_1 = u_1' + u_{10} \geq 0$ and $u_2 = u_2' + u_{20} \geq 0$ because u_1 and u_2 are physical flow rates. Then,

$$\dot{S}(x') \leq -(x_1' - x_2')^2 + y_1^* u_1' + y_2^* u_2'$$

$$\leq y^{*T} u'. \tag{2.135}$$

Therefore, the process is passive with respect to the equilibrium $x' = [0,0]^T$. It is interesting to point out that

1. Similar to the linear case, the heat exchanger is inherently passive because the passivity condition is valid for *any* design parameters, types of fluid and operating conditions (different T_{ci} and T_{hi}).
2. The system is passive with respect to any physical equilibrium point $[x_{10}, x_{20}]^T$ because (2.135) holds for any x_0.
3. The equilibrium point x_0 is GS but not GAS. If $x_1' = x_2' \neq 0$, the unforced system does not converge to $x' = 0$.
4. Output rescaling is equivalent to sensor calibration. Because T_{ci} is never greater than T_{co}, the rescaling coefficient for y_1^* is non-positive. A higher inlet cold stream flow rate will lead to a lower outlet temperatures (T_{co} and T_{ho}). This implies that the direction of x' movement has to be reversed to obtain a minimum phase condition.

In addition, as the system outputs are simply rescaled states, the above system is ZSD. As a result, the heat exchanger is very easy to control. According to Proposition 2.50, any output feedback controller (a mapping from y^* to u') which is dissipative with respect to a supply rate of $w = u'^T y^* - \nu y^{*T} y^* - \rho u'^T u'$, $\rho > 0$ and $\nu > 0$ (*i.e.*, with simultaneous excessive IFP and OFP) will asymptotically stabilize the equilibrium $x' = [0, 0]^T$. A special case is a proportional only controller $u' = -ky^*$ for any $k > 0$.

2.8 Summary

In this chapter, the basic concepts of dissipative systems and passive systems are introduced. The input-output properties of passive systems are discussed. These properties lead to useful stability conditions for interconnected systems, on which the developments described in later chapters build. At first glance, it seems that the stability conditions based on passivity could be conservative compared to those based on dissipativity, because passive systems are a special case of dissipative systems. With the notions of IFP and OFP, the conservativeness vanishes because dissipative systems with respect to different supply rates can be represented by passive systems with certain IFP and OFP. Excess and shortage of IFP and OFP are also used to characterize processes in terms of their passivity. In the next few chapters, passivity-based system analysis and control design are developed for linear processes. These approaches can be implemented numerically and applied directly in routine process control practice.

3

Passivity-based Robust Control

In Chapter 2, we have seen that input feedforward passivity (IFP) is a phase related property. Therefore, it is possible to characterize the uncertainties in terms of their IFP so that both the phase and gain information of the uncertainties can be used, leading to a potentially less conservative control design approach. Robust control design methods based on the passivity uncertainty bound are presented in this chapter, along with case studies and illustrative examples. The developments are for linear systems, leading to systematic approaches that can be readily applied to process control practice.

3.1 Introduction

Model based control is very attractive because it provides systematic procedures for controller design and can achieve good control performance. Various control techniques have been proposed with different features. However, all such techniques must face the formidable adversaries of plant variability and uncertainty. For example, the performance of linear quadratic Gaussian (LQG) controllers can be arbitrarily bad, and even the stability of the closed-loop system of the process and controller cannot be guaranteed if there exists model–plant mismatch (also known as uncertainty). Unfortunately, perfect models are very rare and in some cases even impossible to obtain, especially for complex chemical processes. So it is very desirable that a model-based controller can tolerate plant variability and uncertainty — *i.e.*, be robust.

A vast amount of theoretical and applied research has been done on robust process control. Currently, the mainstream of robust process control is based on the Small Gain Theorem (such as $\mathcal{H}_\infty$ control). The basic idea is to quantify the uncertainty in terms of the bound of its gain and design a controller to stabilize the process when the uncertainty is within the gain bound (represented using system norms). A very brief summary of $\mathcal{H}_\infty$ control is presented in this section to lay the foundation for development of the passivity-based

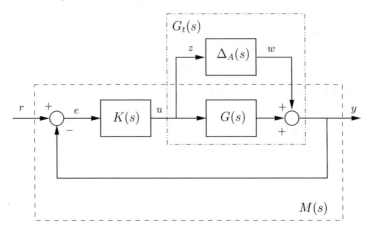

Fig. 3.1. Model with additive uncertainty

approach. Thorough treatments of linear robust control can be found in [46] and [156].

3.1.1 Uncertainties

The uncertainties can be classified into two categories: parametric uncertainty and unstructured uncertainty. In the *parametric uncertainty* case, the structure of the model is known, but certain parameters are unknown. Denote the actual but unknown linear plant as

$$G_t(s) := (A_t, B_t, C_t, D_t), \tag{3.1}$$

where $G_t(s)$ is the transfer function and (A_t, B_t, C_t, D_t) is its state-space representation, and the nominal plant (the model) as

$$G(s) := (A_g, B_g, C_g, D_g). \tag{3.2}$$

The parametric uncertainty represents parametric variations in plant dynamics, *e.g.*, the uncertainties in certain entries of the state-space matrices (A_g, B_g, C_g, D_g) or in specific loop gain and/or poles and/or zeros of the plant transfer function. *Unstructured uncertainty* assumes less knowledge of the process. Model errors include inaccurate or missing dynamics. The most commonly used representations of unstructured uncertainty are

- *Additive Uncertainty.* The simplest way to express the difference between the model and the true system is additive representation (as shown in Figure 3.1):

$$G_t(s) = G(s) + \Delta_A(s), \tag{3.3}$$

  where the model uncertainty is given by

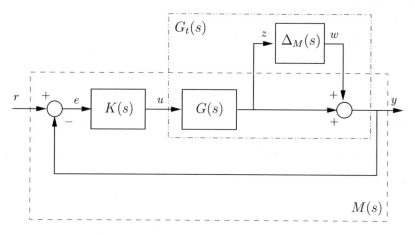

Fig. 3.2. Model with multiplicative uncertainty

$$\Delta_A (s) = G_t (s) - G(s). \tag{3.4}$$

- *Multiplicative Uncertainty.* The model uncertainty may also be represented in the multiplicative form (as shown in Figure 3.2):

$$G_t (s) = [I + \Delta_M (s)] G(s), \tag{3.5}$$

so that $\Delta_M(s)$ is the modelling error relative to the nominal model, where,

$$\Delta_M (s) = [G_t(s) - G(s)] G^{-1}(s). \tag{3.6}$$

Although the uncertainty systems are unknown, their bounds can often be estimated. For example, in $\mathcal{H}_\infty$ control, the largest possible "magnitudes" of the uncertainties are used to estimate how "large" the uncertainties are. The above uncertainties ($\Delta_A (s)$ and $\Delta_M (s)$) are represented as dynamic systems. Therefore, the uncertainty at different frequencies can be represented by frequency responses, *e.g.*, $\Delta_A (j\omega) = G_t (j\omega) - G (j\omega)$. The magnitudes of the uncertainties can be described by using maximum singular values, *e.g.*, $\bar{\sigma} (\Delta_A (j\omega))$. The frequency-dependent description of uncertainties is useful because model uncertainties are often small at low frequencies and increase to unity and above at high frequencies [155]. In this case, the uncertainty regions at different frequencies are assumed to be disc-shaped, $\bar{\sigma} [\Delta_A (j\omega)] < \gamma (\omega)$.

A weighting function can be introduced to normalize the uncertainty to less than 1 in magnitude at all frequencies. For example, as shown in Figures 3.1 and 3.2, stable and rational transfer functions $W_A(s)$ and $W_M(s)$ can be found such that $\Delta_A(s)$ and $\Delta_M(s)$ can be represented as

$$\Delta_A(s) = W_A(s)\Delta(s), \tag{3.7}$$
$$\Delta_M(s) = W_M(s)\Delta(s), \tag{3.8}$$

where $\bar{\sigma}\left(\Delta\left(j\omega\right)\right) < 1$, $\forall\ \omega \in \mathbb{R}$.

Given a transfer function $G(s)$ that is analytic and bounded in the open RHP, its $\mathcal{H}_\infty$-norm is defined as [156]

$$\|G\|_\infty \triangleq \sup_{\operatorname{Re}(s)>0}\ \bar{\sigma}\left[G\left(s\right)\right] = \sup_{\omega \in \mathbb{R}} \bar{\sigma}\left[G\left(j\omega\right)\right]. \qquad (3.9)$$

Assuming that the uncertainty $\Delta(s)$ is stable, then its bound can be neatly represented by the $\mathcal{H}_\infty$ system norm: $\|\Delta\|_\infty < 1$. More generally, the weighting functions can be placed on both the input and output sides of the uncertainty, e.g.,

$$\Delta_A(s) = W_{Ao}(s)\Delta(s)W_{Ai}\left(s\right), \qquad (3.10)$$

$$\Delta_M(s) = W_{Mo}(s)\Delta(s)W_{Mi}\left(s\right). \qquad (3.11)$$

For multivariable systems, the multiplicative uncertainty can be defined in both input and output forms. Uncertainty $\Delta_M(s)$ given in (3.6) is actually the output multiplicative uncertainty because the uncertainty system is on the output side (shown in Figure 3.2). Input multiplicative uncertainty can be defined as follows [156]:

$$\Delta_{MI}\left(s\right) = G^{-1}(s)\left[G_t(s) - G(s)\right]. \qquad (3.12)$$

Another type of representation of uncertainty is called *coprime factor uncertainty* [85], which is particularly suitable for unstable nominal models. In this chapter, we focus on the representations of additive and multiplicative uncertainties, to which passivity-based analysis is more readily applied.

3.1.2 Robust Stability

We can regroup the subsystems in Figures 3.1 or 3.2 such that the closed-loop system can be represented as the feedback system of the uncertainty $\Delta(s)$ (from z to w) and the rest of subsystem $M(s)$ "seen" by the uncertainty (from w to z). The robust stability problem reduces to the stability of the interconnected system in Figure 3.3.

The Small Gain Theorem [147] gives one sufficient condition for robust stability. It can be described as: for the feedback system depicted in Figure 3.3, if both Δ and M are stable and $\|\Delta\|_\infty \|M\|_\infty < 1$, then the closed-loop system is stable. If $\Delta(s)$ is normalised as in (3.7) and (3.8), then the Small Gain Theorem requires that

$$\|M\|_\infty < 1. \qquad (3.13)$$

For additive uncertainty,

$$\begin{aligned} M\left(s\right) &= K(s)\left[I + G\left(s\right)K\left(s\right)\right]^{-1}W_A \\ &= K(s)S\left(s\right)W_A, \end{aligned} \qquad (3.14)$$

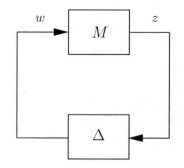

Fig. 3.3. $M - \Delta$ structure for robust stability analysis

where

$$S(s) = [I + G(s)K(s)]^{-1} \tag{3.15}$$

denotes the *sensitivity function*.

For multiplicative uncertainty,

$$\begin{aligned} M(s) &= G(s)K(s)[I + G(s)K(s)]^{-1}W_M \\ &= T(s)W_M, \end{aligned} \tag{3.16}$$

where

$$T(s) = G(s)K(s)[I + G(s)K(s)]^{-1} \tag{3.17}$$

denotes the *complementary sensitivity function*.

In $\mathcal{H}_\infty$ control, the controller $K(s)$ is designed such that the small gain condition given in (3.13) is satisfied to guarantee robust stability.

3.2 Characterization of Uncertainties

3.2.1 Uncertainty Bound Based on IFP

As shown in the previous section, in $\mathcal{H}_\infty$ control, the uncertainty is characterized entirely by its gain, *e.g.*, discs in a complex domain. To use the phase information on the uncertainty, we can characterize the uncertainties with their input feedforward passivity index. Assume negative feedback in Figure 3.3. If the uncertainty is stable and passive with IFP ($\nu > 0$), then a controller can be designed such that the closed-loop system $M(s)$ is stable and strictly input passive to achieve robust stability. If the uncertainty has a shortage of IFP ($\nu < 0$), then the closed-loop system $M(s)$ has to have excessive OFP ($\rho > -\nu$) to ensure robustness. This condition implies that system $M(s)$ has limited gain.

The frequency-dependent IFP index, as given in (2.90), can be adopted to characterize the uncertainty at different frequencies. In this case, the input feedforward is a dynamic system. The IFP index contains both the

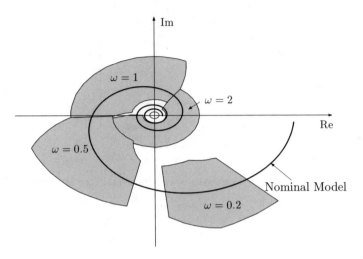

Fig. 3.4. Uncertainty regions

gain and phase information on the uncertainty. This can be seen by a single input single output (SISO) example. For a SISO uncertainty system $\Delta(s)$, its IFP index is reduced to the real part of its frequency response $\nu_F(\Delta(s),\omega) = \mathrm{Re}(\Delta(j\omega))$. Represent $\Delta(j\omega) = r(\omega)[\cos\theta(\omega) + j\sin\theta(\omega)]$, where $r(\omega)$ is the gain and $\theta(\omega)$ is the phase of the uncertainty. Then the IFP index $\nu_F(\Delta(s),\omega) = \mathrm{Re}(\Delta(j\omega)) = r(\omega)\cos\theta(\omega)$.

Because we are concerned only with the shortage of IFP of the uncertainty, the following uncertainty bound can be defined.

Definition 3.1 (Passivity-based uncertainty measure). *The passivity-based uncertainty measure is defined as*

$$
\begin{aligned}
\nu_-(\Delta(s),\omega) &\triangleq -\nu_F(\Delta(s),\omega) \\
&= -\frac{1}{2}\underline{\lambda}(\Delta(j\omega) + \Delta^*(j\omega)),
\end{aligned}
\tag{3.18}
$$

where $\underline{\lambda}$ denotes the minimum eigenvalue.

The above uncertainty measure indicates the *worst case shortage of input passivity*. To illustrate further the property of this passivity-based uncertainty measure, let us consider the following set of uncertain plants.

Example 3.2. Assume that Π is the set of uncertain plants generated by varying the parameters of the following transfer function:

$$
G_t(s) = \frac{k}{\tau s + 1} e^{-\theta s}, \quad 2 \le k, \theta, \tau \le 4.
\tag{3.19}
$$

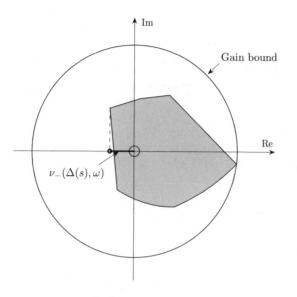

Fig. 3.5. Passivity-based uncertainty bound

The nominal plant model is

$$G\left(s\right) = \frac{3}{3.5s + 1}e^{-3.5}. \tag{3.20}$$

The Nyquist regions for $G_t\left(s\right) \in \Pi$ are generated as shown in Figure 3.4. In particular, the frequency response of the additive uncertainty $\Delta_A^{'}\left(j\omega\right) = G_t\left(j\omega\right) - G\left(j\omega\right)$ at $\omega = 0.2$ rad/s is shown in Figure 3.5.

From the above example, it can be seen that the passivity-based uncertainty bound, $\nu_-\left(\Delta\left(s\right), \omega\right)$, is never greater than the gain bound. In many cases, such as Example 3.2, the frequency response of the nominal model is not situated in the centre of the Nyquist region, therefore, the IFP index bound can be significantly smaller than that of the gain bound, leading to potentially less conservative robust control.

For more general multivariable uncertainties, the passivity-based uncertainty bound has the following property:

Property 3.3 (IFP Index). For any stable multivariable linear system $\Delta\left(s\right)$,

1. $\nu_-\left(\Delta(s), \omega\right) \leq \bar{\sigma}\left(\Delta(j\omega)\right)$;
2. If $\Delta(s) = \Delta_p(s) + \Delta_{np}(s)$, where Δ_p is passive and Δ_{np} is non-passive then,

$$\nu_-\left(\Delta(s), \omega\right) \leq \bar{\sigma}\left(\Delta_{np}(j\omega)\right).$$

Proof.

1. Since $|\lambda[\Delta(j\omega)]| \leq \bar{\sigma}[\Delta(j\omega)]$,

$$
\begin{aligned}
\underline{\lambda}\left(\Delta(j\omega) + \Delta^*(j\omega)\right) &\geq -\bar{\sigma}\left(\Delta(j\omega) + \Delta^*(j\omega)\right) \\
&\geq -\bar{\sigma}\left(\Delta(j\omega)\right) - \bar{\sigma}\left(\Delta^*(j\omega)\right) \qquad (3.21) \\
&\geq -2\bar{\sigma}\left(\Delta(j\omega)\right).
\end{aligned}
$$

Therefore,

$$
\begin{aligned}
\nu_-\left(\Delta(s),\omega\right) &\triangleq -\frac{1}{2}\underline{\lambda}\left(\Delta(j\omega) + \Delta^*(j\omega)\right) \\
&\leq \bar{\sigma}[\Delta(j\omega)], \quad \forall \, \omega \in \mathbb{R}.
\end{aligned} \qquad (3.22)
$$

2. From the Weyl inequality [4],

$$
\begin{aligned}
\underline{\lambda}&\left(\frac{1}{2}[\Delta(j\omega) + \Delta^*(j\omega)]\right) \\
&\geq \underline{\lambda}\left(\frac{1}{2}[\Delta_p(j\omega) + \Delta_p^*(j\omega)]\right) + \underline{\lambda}\left(\frac{1}{2}[\Delta_{np}(j\omega) + \Delta_{np}^*(j\omega)]\right) \qquad (3.23) \\
&\geq \frac{1}{2}\left(\underline{\lambda}\left(\Delta_p(j\omega) + \Delta_p^*(j\omega)\right) - \bar{\sigma}\left(\Delta_{np}(j\omega) + \Delta_{np}^*(j\omega)\right)\right) \\
&\geq -\bar{\sigma}\left(\Delta_{np}(j\omega)\right).
\end{aligned}
$$

Therefore,

$$
\nu_-\left(\Delta(s),\omega\right) \leq \bar{\sigma}\left(\Delta_{np}(j\omega)\right). \qquad (3.24)
$$

Based on the above frequency-dependent passivity-based uncertainty measure, the following robust stability condition can be obtained:

Proposition 3.4. *Consider the feedback system shown in Figure 3.3, where $M(s)$ and $\Delta(s)$ are linear time invariant (LTI) systems. Assume that the uncertainty $\Delta(s)$ is stable with $\nu_-(\Delta(s),\omega) \leq \nu_F(W(s),\omega)$, where $W(s)$ is a stable and minimum phase transfer function. The closed-loop system is stable if*

$$
M'(s) = M(s)[I - W(s)M(s)]^{-1} \qquad (3.25)
$$

is stable and strictly input feedforward passive.

Because for linear systems, the term *strictly passive systems* is a synonym of *stable and strictly input passive systems*, we will use the term *strictly passive systems* from this point. The proof of the above condition is straightforward. It is obvious that $\Delta(s) + W(s)$ is positive real, because $\nu_-(\Delta(s),\omega) < \nu_F(W(s),\omega)$. Perform loop shifting as shown in Figure 3.6. According to Proposition 2.47, it is obvious that $M(s)'$, which is $M(s)$ with positive feedback of $W(s)$, needs to be strictly passive to ensure closed-loop stability. This is a generalized excessive OFP condition on $M(s)$, where feedback is a dynamic system.

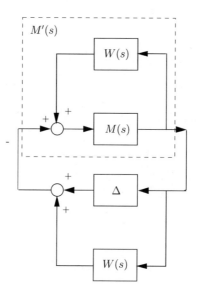

Fig. 3.6. Loop shifting

3.2.2 Uncertainty Bounds Based on Simultaneous IFP and OFP

Proposition 3.4 gives the condition of robust stability for any stable uncertainty bounded only by the passivity-based IFP measure, including uncertainties with very large gain. This may lead to stringent conditions on system $M(s)$:

- $M(s)$ must have excessive OFP and must be minimum phase. For example, if $\Delta(s)$ is a multiplicative uncertainty, then,

$$M(s) = T(s) = G(s) K(s) [I + G(s) K(s)]^{-1}. \qquad (3.26)$$

If the process $G(s)$ is not minimum phase, then it is not possible to find a controller $K(s)$ such that $M(s)$ is minimum phase.
- In robust control design based on the above condition, a controller $K(s)$ must be found to render the closed-loop system $M'(s)$ strictly passive (or strictly positive real). An explicit solution to this problem can be difficult because $M'(s)$ may have a relative degree of 1 which may lead to numerical problems in the positive-real lemma (see Lemma 2.16). The practical approach is to render $M'(s)$ extended strictly positive real (ESPR) (see Definition 2.24) [102, 123]. However, if $G(s)$ is strictly proper, then a proper control $K(s)$ cannot render $M'(s)$ ESPR.

To remove the above restrictions, we can characterize the uncertainty by using both IFP and OFP indices. If the uncertainty $\Delta(s)$ is stable and gain

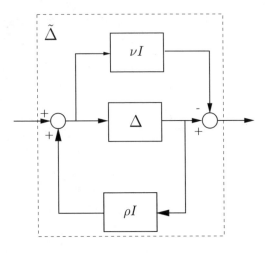

Fig. 3.7. Uncertainty measure with simultaneous IFP and OFP

bounded, then there exists a $\rho > 0$ such that the $\Delta(s)$ with simultaneous negative feedforward νI and positive feedback of ρI is passive (as shown in Figure 3.7). Because $\Delta(s)$ has excessive OFP, according to Theorem 2.49, system $M(s)$ (and also system $M'(s)$) is only required to be IFP(ν_M) such that $\nu_M > -\rho$. Neither $M(s)$ nor $M'(s)$ needs to be ESPR. Such a simultaneous IFP and OFP bound can be more conveniently represented using the concept of a system sector.

Definition 3.5 (Sector of a system [102]). *A stable linear system $T(s)$ is said to be inside the sector $[a, b]$, where a and b are real numbers with $a < b$, $0 < b < +\infty$, if*

$$\mathrm{Re}\left(\left[I - b^{-1}T(s)\right]^* [T(s) - aI] \right) \geq 0, \ \forall \ s = j\omega \tag{3.27}$$

and $T(s)$ is said to be outside the sector $[a, b]$ if

$$\mathrm{Re}\left(\left[I - b^{-1}T(s)\right]^* [T(s) - aI] \right) < 0, \ \forall \ s = j\omega, \tag{3.28}$$

where

$$\mathrm{Re}\left(T(j\omega)\right) \triangleq \frac{1}{2}\left[T(j\omega) + T^*(j\omega)\right]. \tag{3.29}$$

The system sector is related to both the passivity bound and the gain bound [44]. For any stable system $T(s)$,

1. $T(s)$ is passive $\iff T(s)$ is inside $[0, +\infty]$.
2. $T(s)$ is strictly passive $\iff T(s)$ is inside $[\delta, \ +\infty - \zeta]$ (where δ and ζ are arbitrarily small positive real numbers).

3. $T(s)$ is IFP(ν) $\Longleftrightarrow$ $T(s)$ is inside $[\nu, +\infty]$.
4. $\|T(s)\|_\infty \leq \gamma \Longleftrightarrow T(s)$ is confined to a symmetrical sector $[-\gamma, +\gamma]$.
5. $T(s)$ with simultaneous negative feedforward νI and positive feedback of ρI is passive $\Longleftrightarrow$ $T(s)$ is inside $[\nu, 1/\rho]$.

Property 3.6 (Properties of Sectors [147]). If a linear system T is inside sector $[a, b]$ and a linear system T_1 is inside $[a_1, b_1]$, then,

1. System kT is inside $[ka, kb]$ (where k is a constant).
2. System $T + T_1$ is inside $[a + a_1, \ b + b_1]$.

Theorem 3.7 (Sector stability theorem [147]). *Consider linear systems G_1 and G_2 in negative output feedback connection. Let them both be confined to sectors. Let ζ and δ be constants, of which one is strictly positive and one is zero. Suppose that G_2 is inside the sector $[a+\zeta, \ b-\zeta]$ $(b > 0)$. The closed-loop system of G_1 and G_2 is stable if G_1 satisfies one of the following conditions:*

1. *If $a > 0$ and G_1 is outside $[-1/a - \delta, \ -1/b + \delta]$.*
2. *If $a < 0$ and G_1 is inside $[-1/b + \delta, \ -1/a - \delta]$.*
3. *If $a = 0$ and G_1 is inside $[-1/b + \delta, \ +\infty]$.*

The above condition is fundamentally equivalent to the stability condition based on simultaneous IFP and OFP (as given in Theorem 2.49). It can also be seen that both the Small Gain Theorem and the Passivity Theorem are special cases of sector stability. A stable system $G(s)$ can be confined in different sectors, for example, $[\nu, +\infty]$, or $[-\gamma, \gamma]$, etc., where $\nu = \frac{1}{2} \min_{\omega \in \mathbb{R}} \underline{\lambda}(G(j\omega) + G^*(j\omega))$, $\gamma = \max_{\omega \in \mathbb{R}} \bar{\sigma}(G(j\omega))$, as shown in Figure 3.8. The lower and upper sector bounds are correlated. A smaller lower bound implies a smaller corresponding upper bound, and vice versa. If the uncertainty system $\Delta(s)$ is confined in the sector $[a, b]$ (where $b \gg |a| \geq 0 \geq a$), to keep closed-loop stability, $T(s)$ only needs to be within the sector of $[-1/b + \sigma, -1/a - \zeta]$ (where σ and ζ can be any positive constants with arbitrarily small values), not necessarily strictly passive. The passivity-based uncertainty measure can be represented as both the lower sector bound (relating to IFP) and upper sector bound (relating to OFP) of the uncertainty system $\Delta(s)$, preferably frequency-dependent so that system robustness and performance can be optimized at different frequencies. However, implementation of both frequency-dependent lower and upper bounds is very complicated. A simplified approach is to characterize the uncertainty using a frequency-dependent lower sector bound with a constant upper bound. This is equivalent to a frequency-dependent IFP index with a constant OFP index:

Definition 3.8 (Sector-bounded passivity uncertainty measure[10]). *Given a stable uncertainty system $\Delta(s)$, for a constant $b \gg |\nu(\Delta(s))|$, the sector-bounded passivity uncertainty measure is defined as*

$$\nu_{S-}(\Delta(s), \omega, b) \triangleq -\sup\left\{ a : \mathrm{Re}\left(\left[I - b^{-1}\Delta(j\omega) \right]^* [\Delta(j\omega) - aI] \right) \geq 0 \right\}.$$
$$(3.30)$$

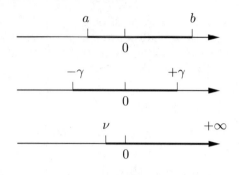

Fig. 3.8. System confined in different sectors

The above measure indicates the minimum feedforward *required* (similar to the ν_- index), such that the uncertainty with *positive* feedback of $1/b$ is passive. The sector-bounded passivity index can be estimated from the following conditions:

Theorem 3.9 ([8]). *For a given stable system $T(s)$ and the upper sector bound b, the sector-bounded passivity index is bounded by the following inequality:*

$$\nu_{S_-}(T(s), \omega, b) \leq \frac{2\bar{\sigma}^2\left(T(j\omega)\right) - b\underline{\lambda}\left(T(j\omega) + T^*(j\omega)\right)}{2b - \underline{\lambda}\left(T(j\omega) + T^*(j\omega)\right)}. \tag{3.31}$$

Proof. Let:

$$a = -\frac{2\bar{\sigma}^2\left(T(j\omega)\right) - b\underline{\lambda}\left(T(j\omega) + T^*(j\omega)\right)}{2b - \underline{\lambda}\left(T(j\omega) + T^*(j\omega)\right)} \tag{3.32}$$

and

$$N = \mathrm{Re}\left(\left[I - \frac{T(j\omega)}{b}\right]^* [T(j\omega) - aI]\right)$$
$$= \left(1 + \frac{a}{b}\right)[T(j\omega) + T^*(j\omega)] - \frac{2T^*(j\omega)T(j\omega)}{b} - 2aI. \tag{3.33}$$

Since N is a Hermitian matrix, the necessary and sufficient condition for N to be positive semidefinite is that its eigenvalues are all nonnegative, *i.e.*, $\underline{\lambda}(N) \geq 0$. Denote $\lambda_m(\omega) = \underline{\lambda}(T(j\omega) + T^*(j\omega))$. Because $T(j\omega) + T^*(j\omega)$, $-2T^*(j\omega)T(j\omega)/b$ and $2aI$ are all Hermitian matrices, from the Weyl inequality [4],

$$\underline{\lambda}(N) \geq \lambda_m(\omega) + \frac{a}{b}\lambda_m(\omega) - 2a - \frac{2\bar{\sigma}^2(T(j\omega))}{b}$$

$$= \lambda_m(\omega) + \frac{2\bar{\sigma}^2(T(j\omega)) - b\lambda_m(\omega)}{\lambda_m(\omega) - 2b}\left[\frac{1}{b}\lambda_m(\omega) - 2\right] - \frac{2\bar{\sigma}^2(T(j\omega))}{b}$$

$$= \lambda_m(\omega) + \frac{2\bar{\sigma}^2(T(j\omega))}{b} - \lambda_m(\omega) - \frac{2\bar{\sigma}^2(T(j\omega))}{b}$$

$$= 0.$$

$$(3.34)$$

Therefore, matrix N is positive semidefinite. From Definition 3.5, $T(j\omega)$ is in the sector $[a, b]$. This proves (3.31).

The sector-bounded passivity measure has the following properties:

Property 3.10 (Properties of sector-bounded passivity index [10]).

1. $\nu_{S-}(T_1(s) + T_2(s), \omega, b_1 + b_2) \leq \nu_{S-}(T_1(s), \omega, b_1) + \nu_{S-}(T_2(s), \omega, b_2)$.
2. For any $\|T(s)\|_\infty \leq b < \infty$, $\nu_{S-}(T(s), \omega, b) \leq \bar{\sigma}(T(j\omega))$.
3. If $T(s)_{m \times m} = \mathrm{diag}\{t_1(s), \ldots, t_i(s), \ldots, t_m(s)\}$, then

$$\nu_{S-}(T(s), \omega, b) \leq \bar{\sigma}(T(j\omega)). \tag{3.35}$$

Proof.

1. This property can be concluded directly from Property 3.6.
2. Denote $\underline{\lambda}(\omega) = \underline{\lambda}(T(j\omega) + T^*(j\omega))$. Since $b \geq \|T(s)\|_\infty$ and $b \geq \frac{1}{2}\underline{\lambda}(\omega)$ for any $\omega \in \mathbb{R}$, from Theorem 3.9,

$$\nu_{S-}(T(s), \omega, b) \leq \frac{2\bar{\sigma}^2(T(j\omega)) - b\underline{\lambda}(\omega)}{2b - \underline{\lambda}(\omega)}$$

$$= \bar{\sigma}(T(j\omega)) + \frac{\{\bar{\sigma}(T(j\omega)) - b\}\{2\bar{\sigma}(T(j\omega)) + \underline{\lambda}(\omega)\}}{2b - \underline{\lambda}(\omega)}.$$

$$(3.36)$$

For any ω,

$$\bar{\sigma}(T(j\omega)) \geq 0, \quad \bar{\sigma}(T(j\omega)) \geq \underline{\lambda}(\omega)/2, \quad b \geq \sigma_{\max}(T(j\omega)). \tag{3.37}$$

This leads to

$$\frac{[\bar{\sigma}(T(j\omega)) - b][2\bar{\sigma}(T(j\omega)) + \underline{\lambda}(\omega)]}{2b - \underline{\lambda}(\omega)} \leq 0. \tag{3.38}$$

Therefore,

$$\nu_{S-}(T(s), \omega, b) \leq \bar{\sigma}(T(j\omega)). \tag{3.39}$$

3. Define:

$$N(\omega) = \mathrm{Re}\left(\left[I - \frac{T(j\omega)}{b}\right]^* [T(j\omega) + \nu_{S-}(T(j\omega),\omega,b)I]\right). \tag{3.40}$$

Because $T(s)$ is diagonal, so is $N(\omega) = \mathrm{diag}\{n_i(\omega)\}$ $(i = 1,\ldots,m)$, where

$$n_i(\omega) = \mathrm{Re}\left([1 - t_i(j\omega)/b]^* [t_i(j\omega) + \nu_{S-}(T(j\omega),\omega,b)]\right), \quad i = 1,\ldots,m. \tag{3.41}$$

The sector-bounded passivity index $\nu_{S-}(T(s),\omega,b)$ is the minimum value such that $N(\omega)$ is positive semidefinite for all ω, that is, $n_i(\omega) \geq 0$, $\forall\,\omega$, $i = 1,\ldots,m$. Define

$$n_i^*(\omega) = \mathrm{Re}\left([1 - t_i(j\omega)/b]^* [t_i(j\omega) + a_i]\right), \quad i = 1,\ldots,m. \tag{3.42}$$

where $a_i = \nu_{S-}(t_i(s),\omega,b)$ is the minimum value such that $n_i^*(\omega) \geq 0$; then

$$\max_i a_i = \max_i \nu_{S-}(t_i(s),\omega,b) \tag{3.43}$$

is the minimum value such that all the inequalities $n_i(\omega) \geq 0$ hold for all $i = 1,\ldots,m$. Therefore, (3.35) holds.

Similar to the IFP index ν_-, the sector-based passivity index also comprises both the phase and gain information of the uncertainty, although it is less obvious.

3.3 Passivity-based Robust Control Framework

Based on the above passivity uncertainty measure, we can develop a passivity-based robust control (PBRC) framework, including robust stability conditions and a control design method, as detailed in [8].

3.3.1 Robust Stability Condition

The robust stability condition on the basis of passivity can be described as follows:

Theorem 3.11 (Passivity-based robust stability condition). *Consider an interconnected system (as shown in Figure 3.6) comprised of $M(s)$ and a stable uncertainty system $\Delta(s)$. Given a stable and minimum phase weighting function $W(s)$ whose sector-bounded passivity index*

$$\nu_{S-}(W(s),\omega,b_w) \leq -\nu_{S-}(\Delta(s),\omega,b) \tag{3.44}$$

the closed-loop system will be stable if

$$M'(s) + \frac{1}{b + b_w}I \tag{3.45}$$

is strictly passive or ESPR, where $M'(s)$ is defined in (3.25).

Proof. Assume $\nu_{S-}(\Delta, \omega, b) = \alpha(\omega)$, where $\alpha(\omega)$ is a nonnegative real function of frequency. Thus $\Delta(s)$ and $W(s)$ are confined in the sectors $[-\alpha(\omega), b]$ and $[\alpha(\omega), b_w]$, respectively. From Property 3.6, system $\Delta(s) + W(s)$ is inside sector $[0, b + b_w]$. From Theorem 3.7, the sufficient condition for the closed-loop system of $M - \Delta$ to be stable is that System $M'(s)$ is in the sector of $\left[-\frac{1}{b+b_w} + \delta, +\infty - \delta\right]$ (where δ is an arbitrarily small positive real number). Again from Property 3.6, $M'(s) + \frac{1}{b+b_w}I$ needs to be confined to $[\delta, +\infty - \delta]$.

The PBRC problem is now formulated as follows: Given a plant model with uncertainty which is bounded by its sector-bounded passivity index, find a controller such that the stability condition in Theorem 3.11 is satisfied. Theorem 3.11 reduces to Proposition 3.4 when the upper bound b equals $+\infty$. Since b is always chosen to be much larger than the system's infinity norm, Theorem 3.11 can be seen as a weak passivity stability condition. Although the sector-bounded passivity index of an uncertainty system is very close to its IFP passivity index when a large upper bound is adopted, the restriction on the relative degree of $M'(s)$ is removed as long as $b < +\infty$.

3.3.2 Robust Stability and Nominal Performance

It is known that robustness and nominal performance compete with each other in $\mathcal{H}_\infty$ control. This is also true in passivity-based robust control. The robust stability condition presented in the previous subsection implies gain constraints on the closed-loop system "seen" by the uncertainty.

Proposition 3.12. *The robust stability condition in Theorem 3.11 implies that for all frequencies,*

$$\bar{\sigma}\left(M(j\omega) - \frac{1}{2}\left[\frac{1}{\nu_{S-}(\Delta(s), \omega, b)} - \frac{1}{b}\right]\right) < \frac{1}{2}\left[\frac{1}{\nu_{S-}(\Delta(s), \omega, b)} + \frac{1}{b}\right].$$
$$(3.46)$$

Proof. Denote $a(\omega) = \nu_{S-}(\Delta(s), \omega, b)$. Define a complex function

$$N(\omega) = M(j\omega)[I - a(\omega)M(j\omega)]^{-1} + \frac{1}{b + a(\omega)}I. \qquad (3.47)$$

Theorem 3.11 requires that $N(\omega) + N^*(\omega) > 0$, $\forall \omega \in \mathbb{R}$. Furthermore,

$$N(\omega) + N^*(\omega)$$
$$= \frac{b}{b + a(\omega)}\left\{\left[M(j\omega) + \frac{1}{b}I\right][I - a(\omega)M(j\omega)]^{-1}\right.$$
$$\left. + [I - a(\omega)M^*(j\omega)]^{-1}\left[M^*(j\omega) + \frac{1}{b}I\right]\right\}$$
$$= \frac{b}{b + a(\omega)}[I - a(\omega)M^*(j\omega)]^{-1}\left\{[I - a(\omega)M^*(j\omega)]\left[M(j\omega) + \frac{1}{b}I\right]\right.$$
$$\left. + \left[M^*(j\omega) + \frac{1}{b}I\right][I - a(\omega)M(j\omega)][I - a(\omega)M(j\omega)]^{-1}\right\}.$$
$$(3.48)$$

Since $\frac{b}{b+a(\omega)} > 0$, the above condition implies for all $\omega \in \mathbb{R}$ that

$$
\underline{\lambda}\left([I - a\left(\omega\right) M^*(j\omega)] \left[M(j\omega) + \frac{1}{b}I \right] \right.
$$
$$
\left. + \left[M^*(j\omega) + \frac{1}{b}I \right] [I - a\left(\omega\right) M(j\omega)] \right) > 0. \quad (3.49)
$$

The above inequality can be rewritten as follows:

$$
\bar{\lambda}\left(2a\left(\omega\right) M^*(j\omega) M(j\omega) + \frac{a\left(\omega\right)}{b} M(j\omega) \right.
$$
$$
\left. + \frac{a\left(\omega\right)}{b} M^*(j\omega) - M(j\omega) - M^*(j\omega) - \frac{2}{b}I \right) < 0. \quad (3.50)
$$

where $\bar{\lambda}$ denote the maximum eigenvalue. After factorization of the left-hand side, for all $\omega \in \mathbb{R}$,

$$
\bar{\lambda}\left\{ \left[M(j\omega) - \frac{1}{2}\left(\frac{1}{a\left(\omega\right)} - \frac{1}{b} \right)I \right]^* \left[M(j\omega) - \frac{1}{2}\left(\frac{1}{a\left(\omega\right)} - \frac{1}{b} \right)I \right] \right\}
$$
$$
< \frac{1}{4}\left(\frac{1}{a\left(\omega\right)} + \frac{1}{b} \right)^2, \quad (3.51)
$$

leading to

$$
\bar{\sigma}\left(M(j\omega) - \frac{1}{2}\left(\frac{1}{a\left(\omega\right)} - \frac{1}{b} \right)I \right) < \frac{1}{2}\left(\frac{1}{a\left(\omega\right)} + \frac{1}{b} \right). \quad (3.52)
$$

   In the feedback control shown in Figures 3.1 and 3.2, one of the common control objectives is to keep the error between the plant output y and the reference r small when disturbance d exists. Therefore, a small sensitivity function $S(s)$ (as defined in (3.15)) is desirable. If the uncertainty is modelled as multiplicative uncertainty, the system $M(s)$ "seen" by the uncertainty is the complementary sensitivity function $T(s)$, defined in (3.17). From the above condition, it is concluded that the gain of $T(s)$ has to be limited by the ν_{S-} index of the uncertainty. This implies the limitation on nominal performance: If at certain frequency ω, $\nu_{S-}\left(\Delta(s), \omega, b \right) \leq a$, the gain constraint implies that it is possible to reduce $\bar{\sigma}\left(S\left(j\omega\right) \right) = \bar{\sigma}\left(I - T\left(j\omega\right) \right)$ to be arbitrarily small for $\frac{1}{a} + \frac{1}{b} \geq 1$ but when $\frac{1}{a} + \frac{1}{b} < 1$, the smallest gain of the sensitivity function is

$$
\underline{\sigma}\left(S\left(j\omega\right) \right) \geq \underline{\sigma}\left(I \right) - \bar{\sigma}\left(T\left(j\omega\right) \right) = 1 - \left(\frac{1}{a} + \frac{1}{b} \right), \quad (3.53)
$$

where $\underline{\sigma}$ denotes the minimum singular value. Because b is usually a large number, practically only the value of a plays a significant role in the above equation.

3.3.3 Advantages and Limitations of Passivity-based Robust Control

Both the Small Gain Theorem and the Passivity Theorem are special cases of sector stability. Given the uncertainty system $\Delta(s)$ inside the sector $[-a, b]$, if a is close to b, then the small-gain based control is not conservative. If a is much less than b, especially if $a \leq 0$, then the passivity-based condition can be significantly less conservative. Because the passivity index of the uncertainty combines both gain and phase information, PBRC will be preferable for uncertain systems with variations small in phase but large in gain.

The main advantage of the PBRC design method compared with small gain based control is that the passivity-based approach is often less conservative. The passivity index of the uncertainty is less than, and sometimes much less than its gain bound. According to the discussion in Section 3.3.2, the bandwidth of the open-loop system subject to closed-loop stability achieved by PBRC can be larger than that of $\mathcal{H}_\infty$ control. This means faster command response and disturbance rejection. Furthermore, if the nominal process model has RHP poles and/or zeros, then it is inevitable that system $M(s)$ (e.g., the complementary sensitivity function $T(s)$ in the case of multiplicative uncertainty) will have its gain rising to its peak before it rolls off [40]. This implies poor robustness at the midfrequency. PBRC does not require the gain condition of $M(s)$ and thus avoids this problem.

The proposed control is limited to control applications where the passivity indices of the uncertainties can be estimated. Typical applications include (1) linear regulatory controller synthesis for nonlinear processes using the robust control framework, where the linearised model at the nominal operating point is used as the control model and the model mismatch between the control model and the linearised models at different operating points is treated as uncertainty; (2) control of processes with unknown parameter variations; (3) control of processes whose uncertainty is passive (e.g., [50]).

3.3.4 Robust Control Design

In this section, we discuss the PBRC design approach. Here we assume that the multiplicative uncertainty representation is used. A similar approach can be derived for other types of uncertainty representation. The robust control synthesis problem is formulated as follows:

Problem 3.13. Given a nominal plant model: $G(s) := (A_g, B_g, C_g, D_g)$ with multiplicative uncertainty $\Delta(s)$ bounded by its sector-bounded passivity index:

$$\nu_{S-}(\Delta(s), \omega, b) \leq -\nu_{S-}(W_2(s), \omega, b_w), \tag{3.54}$$

where $W_2(s)$ is a stable and minimum phase weighting function. Design a controller $K(s)$ to stabilize the plant robustly and achieve the following nominal performance:

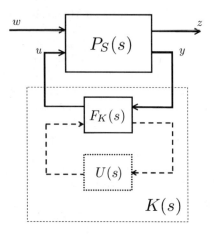

Fig. 3.9. All solution controller for performance

$$\|W_1 S\|_\infty < 1, \tag{3.55}$$

and

$$T(s)[I - W_2(s)T(s)]^{-1} + \frac{1}{b + b_w}I \quad \text{is ESPR.} \tag{3.56}$$

Where weighting function $W_1(s)$ is used to specify required nominal performance. $S(s)$ and $T(s)$ are the sensitivity and complementary functions, respectively (as given in (3.15) and (3.17)). .

The challenge is to find a controller such that both conditions in (3.55) and (3.56) are satisfied. Because the passivity and $\mathcal{H}_\infty$ conditions are required for entirely different closed-loop systems as given in (3.56) and (3.55), existing multi-objective controller synthesis methods such as those used in mixed $\mathcal{H}_2/\mathcal{H}_\infty$ output feedback synthesis [19, 68] cannot be used to solve the proposed problem. A two-step approach is developed as follows. The basic idea is first to find all the controllers that satisfy (3.56) and then select the suitable final controller that also satisfies (3.55).

Step 1. Find the all-solution controller for the nominal performance specification

The problem of finding a controller $K(s)$ that satisfies the constraint in (3.55) alone is that it may lead to controllers with infinite gain. To get sensible solutions, a constant weight w_3 is used to constrain the gain of the controller indirectly. Assume that $W_1(s) := (A_{w1}, B_{w1}, C_{w1}, D_{w1})$. The performance specification (3.55) is revised as follows:

$$\|P_S\|_\infty < 1, \tag{3.57}$$

where

$$P_S\left(s\right) = \begin{bmatrix} W_1(s)S(s) \\ w_3 K\left(s\right) S\left(s\right) \end{bmatrix} \tag{3.58}$$

has the following state-space representation:

$$P_S\left(s\right) := \left[\begin{array}{ccc|c} A_g & 0 & 0 & B_g \\ -B_{w1}C_g & A_{w1} & B_{w1} & -B_{w1}D_g \\ -D_{w1}C_g & C_{w1} & D_{w1} & -D_{w1}D_g \\ 0 & 0 & 0 & w_3 I \\ \hline -C_g & 0 & I & -D_g \end{array}\right]. \tag{3.59}$$

By using an $\mathcal{H}_\infty$ control synthesis tool (e.g., [35]), the all-solution controller

$$F_K(s) = \begin{bmatrix} F_{11}(s) & F_{12}(s) \\ F_{21}(s) & F_{22}(s) \end{bmatrix} := \left[\begin{array}{c|cc} A_F & B_{F1} & B_{F2} \\ \hline C_{F1} & D_{F11} & D_{F12} \\ C_{F2} & D_{F21} & D_{F22} \end{array}\right] \tag{3.60}$$

can be found. As shown in Figure 3.9, $F_K(s)$ is a two-port system such that all controllers $K(s)$ that satisfy (3.57) can be parameterized by a stable contraction map $U(s)$ (i.e., $\|U\|_\infty \le 1$):

$$K(s) = F_{11}(s) + F_{12}(s)U(s)[I - F_{22}(s)U(s)]^{-1}F_{21}(s). \tag{3.61}$$

The right-hand side of the above equation is called the lower linear fractional transformation of $F_K(s)$ and $U\left(s\right)$, denoted as $\mathcal{F}_l(F_K(s), U(s))$.

Step 2. Determine the contraction map $U(s)$ such that $K(s)$ also meets the ESPR Condition

Assume that $W_2(s) := (A_{w2}, B_{w2}, C_{w2}, D_{w2})$. An augmented plant model $P_P(s)$ can be derived to include both the process system model and the weighting function W_2. It has the following state-space representation:

$$P_P\left(s\right) := \left[\begin{array}{cc|c|c} A_g & 0 & 0 & B_g \\ B_{w2}C_g & A_{w2} & 0 & B_{w2}D_g \\ -C_g & 0 & \frac{1}{b+b_w}I & -D_g \\ \hline D_{w2}C_g - C_g & C_{w2} & -I & D_{w2}D_g - D_g \end{array}\right]. \tag{3.62}$$

Condition (3.56) becomes $\mathcal{F}_l\left(P_P(s), K(s)\right)$ is ESPR.

The above control problem is converted into an $\mathcal{H}_\infty$ problem by using the following relationship between passivity and the small gain condition:

Theorem 3.14 ([6]). *Consider a linear system with a transfer function $T(s)$. Denote*

$$T'(s) = [\zeta I - T(s)] [\zeta I + T(s)]^{-1}, \tag{3.63}$$

where ζ can be any positive real number. System $T(s)$ is ESPR if and only if $T'(s)$ is stable and $\|T'\|_\infty < 1$.

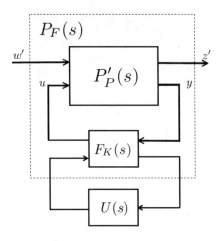

Fig. 3.10. Control design for robustness

The transformation given in (3.63) is called *the Cayley transformation*. By using (3.63), $P_P(s)$ can be converted to $P_P'(s)$ such that $\mathcal{F}_l\left(P_P(s), K(s)\right)$ is ESPR if and only if $\mathcal{F}_l\left(P_P'(s), K(s)\right)$ is stable and

$$\left\|\mathcal{F}_l\left(P_P', K\right)\right\|_\infty < 1. \tag{3.64}$$

$P_P'(s)$ has the following state-space representation:

$$P_P'(s) := \left[\begin{array}{c|c|c} A_P & B_{P1} & B_{P2} \\ \hline C_{P1} & D_{P11} & D_{P12} \\ \hline C_{P2} & D_{P21} & D_{P22} \end{array}\right], \tag{3.65}$$

where

$$A_P = \begin{bmatrix} A_g & 0 \\ B_{w2}C_g & A_{w2} \end{bmatrix},$$

$$B_{P1} = \begin{bmatrix} 0 \\ 0 \end{bmatrix}, \qquad\qquad B_{P2} = \begin{bmatrix} B_g \\ B_{w2}D_g \end{bmatrix},$$

$$C_{P1} = \begin{bmatrix} \frac{2(b+b_w)}{b+b_w+1}C_g & 0 \end{bmatrix}, \quad C_{P2} = \begin{bmatrix} D_{w2}C_g - \frac{2(b+b_w)+1}{b_w+b_w+1}C_g & C_{w2} \end{bmatrix},$$

$$D_{P11} = \frac{b+b_w-1}{b+b_w+1}I, \qquad D_{P12} = \frac{2(b+b_w)}{b+b_w+1}D_g,$$

$$D_{P21} = -\frac{b+b_w}{b+b_w+1}I, \qquad D_{P22} = D_{w2}D_g - \frac{2(b+b_w)+1}{b_w+b_w+1}D_g.$$

Interconnect $P_P'(s)$ with $F_K(s)$ to form another two-port system, as shown in Figure 3.10:

$$P_F(s) = \begin{bmatrix} P_{F11}(s) & P_{F12}(s) \\ P_{F21}(s) & P_{22}(s) \end{bmatrix} = \mathcal{F}_l\left(P'_P(s), F_K(s)\right). \tag{3.66}$$

Now the control design problem becomes

Problem 3.15. To find a *stable* map $U(s)$ such that

$$\|U\|_\infty < 1, \tag{3.67}$$

and

$$\|\mathcal{F}_l(P_F, U)\|_\infty < 1. \tag{3.68}$$

The gain of $U(s)$ should be limited to satisfy Condition (3.67). This can be done by augmenting P_F to

$$P'_F(s) = \begin{bmatrix} \begin{bmatrix} P_{F11}(s) \\ 0 \end{bmatrix} & \begin{bmatrix} P_{F12}(s) \\ w_4 I \end{bmatrix} \\ P_{F21}(s) & P_{F22}(s) \end{bmatrix} := \left[\begin{array}{c|c|c} A_{PF} & B_{PF1} & B_{PF2} \\ \hline C_{PF11} & D_{PF111} & D_{PF121} \\ C_{PF12} & D_{PF112} & D_{PF121} \\ \hline C_{PF2} & D_{PF21} & D_{PF22} \end{array} \right], \tag{3.69}$$

where,

$$A_{PF} = \begin{bmatrix} A_P + B_{P2}D_{F11}XC_{P2} & B_{P2}YC_{F2} \\ B_{F1}XC_{P2} & A_F + B_{F1}XD_{P22}C_{F1} \end{bmatrix},$$

$$B_{PF1} = \begin{bmatrix} B_{P1} + B_{P2}D_{F11}XD_{P21} \\ B_{F1}XD_{P21} \end{bmatrix}, \quad B_{PF2} = \begin{bmatrix} B_{P2}YD_{F12} \\ B_{F2} + B_{F1}D_{P22}YD_{F12} \end{bmatrix},$$

$$C_{PF11} = \begin{bmatrix} C_{P1} + D_{P12}D_{F11}XC_{P2} & D_{P12}YC_{F1} \end{bmatrix}, \quad C_{PF12} = \begin{bmatrix} 0 & 0 \end{bmatrix},$$

$$C_{PF2} = \begin{bmatrix} D_{F21}XC_{P2} & C_{F2} + D_{F21}D_{P22}YC_{F1} \end{bmatrix},$$

$$D_{PF111} = D_{P11} + D_{P12}D_{F11}XD_{P21}, \quad D_{PF121} = D_{P12}YD_{F12},$$

$$D_{PF112} = 0, \quad D_{PF122} = w_4 I,$$

$$D_{PF21} = D_{F21}XD_{P21}, \quad D_{PF22} = D_{F22} + D_{F21}D_{P22}YD_{F12},$$

$$X = (I - D_{P22}D_{F11})^{-1}, \text{ and } Y = (I - D_{F11}D_{P22})^{-1}.$$

By choosing a suitable weighting function w_4, $U(s)$ can be found by solving the following $\mathcal{H}_\infty$ control problem:

$$\|\mathcal{F}_l(P'_F, U)\|_\infty < 1. \tag{3.70}$$

Inequality 3.70 is equivalent to (3.68) and

$$\left\| w_4 (I - U P_{F22})^{-1} U P_{F21} \right\|_\infty < 1. \tag{3.71}$$

Because the gain of $U(s)$ is penalised indirectly, some iterations are necessary to find a suitable $U(s)$ that satisfies $\|U\|_\infty > 1$. A larger w_4 leads to a smaller $\mathcal{H}_\infty$-norm of $U(s)$. A too large w_4 will make Problem (3.70) infeasible.

Control Design Procedure

The control design procedure can be summarized as follows:

Procedure 3.16 (Passivity-based robust control design)

1. *Estimate the passivity index bound of the uncertainty and represent it using weighting function $W_2(s)$.*
2. *Determine the level of control performance that should be achieved and represent it using weighting function $W_1(s)$.*
3. *Choose a small constant for weighting w_3. This value can be arbitrarily small as long as it does not cause numerical problems, e.g., $w_3 = 10^{-4}$.*
4. *Find the augmented plants P_S and P_P as shown in (3.59) and (3.62). Calculate P'_P according to (3.65).*
5. *Find the all-solution controller $F_K(s)$ for the nominal performance specification given in (3.57).*
6. *Interconnect P'_P with $F_K(s)$ to form P_F.*
7. *Solve Problem (3.70) to find the contraction map $U(s)$. Some iterations are necessary to find a suitable $U(s)$. The ∞-norm and stability of $U(s)$ need to be checked in each round of iteration. The initial value of w_4 can be chosen as 1. If $\|U\|_\infty > 1$, then increase w_4 and repeat Step 7; If there is no solution for Problem (3.70), then decrease w_4 and repeat Step 7.*
8. *If no solution can be found for Problem 3.15, then the performance specification is unrealistic. Use a $W_1(s)$ with smaller gain and repeat Steps 3–7.*
9. *Calculate the final controller $K(s) = \mathcal{F}_l(F_K(s), U(s))$.*

Noted that Steps 4–7 and 9 can be coded into a computer program and executed automatically.

Matrix Inequality Approach

An alternative approach is to convert the controller synthesis problem into a feasibility problem with matrix inequality constraints. Condition (3.56) can be cast into a matrix inequality by using the positive-real lemma (Lemma 2.16). Similarly, Condition (3.55) can be represented in a matrix inequality by using the following bounded-real lemma:

Lemma 3.17 (Bounded-real Lemma [21]). *Consider an LTI stable system with a transfer function $G(s) = D + C(sI - A)^{-1} B$. The system norm $\|G\|_\infty < \gamma$ if and only if there exists a symmetrical matrix $P > 0$ such that*

$$\begin{bmatrix} A^T P + P A & P B & C^T \\ B^T P & -\gamma I & D^T \\ C & D & -\gamma I \end{bmatrix} < 0. \tag{3.72}$$

Now the control design problem is cast into a bilinear matrix inequality (BMI) problem. The BMI constraints can be converted into LMIs using a variable transformation technique [107]. The controller can then be designed by using a semidefinite programming (SDP) tool [10].

3.3.5 Example of CSTR Control

The control of an endothermic reaction is used to illustrate the proposed controller design. The reaction tank with a warming jacket is shown in Figure 3.11. The reactant with concentration C_{a0} is fed into the tank with flow rate F. The material reacts and absorbs heat in the tank. Assume that the outlet flow rate is controlled so it is the same as the inlet flow rate. The hot water at temperature T_{j0} is fed into the jacket with flow rate F_j. Constant holdup and perfect mixing are assumed in the tank and the jacket. The control system is designed to control the concentration C_a and temperature T in the tank by manipulating flow rates F and F_j.

The dynamics of the system are described as

$$\frac{dC_a}{dt} = \frac{1}{V}\left[F(C_{a0} - C_a) - V\alpha\exp(-\frac{E}{RT})C_a\right],$$

$$\frac{dT}{dt} = \frac{1}{V}\left[FT_0 - FT - \frac{\lambda V\alpha\exp(-\frac{E}{RT}C_a)}{\rho C_p} - \frac{UA_H(T - T_j)}{\rho C_p}\right], \quad (3.73)$$

$$\frac{dT_j}{dt} = \frac{F_j(T_{j0} - T_j)}{V_j} + \frac{UA_H(T - T_j)}{\rho_j V_j C_j}.$$

The parameters are $V_j = 3.85$ ft^3, $\alpha = 7.08 \times 10^{10}$ h^{-1}, $E = 30 \times 10^3$ Btu/lb·mol, $A_H = 250$ ft^2, $R = 1.99$ Btu/lb·mol, $U = 150$ Btu/h ft^2 °R,

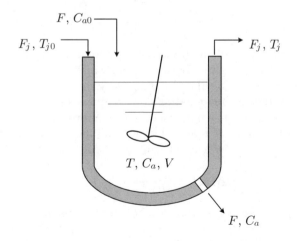

Fig. 3.11. CSTR

$\rho_j = 62.3 \ \mathrm{lb_{mol}/ft^3}, \ \lambda = 30 \times 10^3 \ \mathrm{Btu/lb\cdot mol}, \ \rho = 50 \ \mathrm{lb_{mol}/ft^3}, \ C_p = 0.75 \ \mathrm{Btu/h \ ft^2 {}^\circ R}$, and $C_j = 1.0 \ \mathrm{Btu/lb_{mol} \ {}^\circ R}$.

The normalised linear model at the steady-state operating point ($F_j = 50 \ \mathrm{ft^3/h}$, $T_{j0} = 700 \ {}^\circ R$, $T_j = 605 \ {}^\circ R$, $F = 40 \ \mathrm{ft^3/h}$, $T_0 = 600 \ {}^\circ R$, $T = 597.7 \ {}^\circ R$, $V = 50 \ \mathrm{ft^3}$, $C_{a0} = 0.5 \ \mathrm{lb\cdot mol/ft^3}$ and $C_a = 0.252 \ \mathrm{lb\cdot mol/ft^3}$) is

$$\dot{x} = \begin{bmatrix} -1.588 & -19.863 & 0 \\ -0.266 & -27.500 & 20.263 \\ 0 & 154.319 & -169.332 \end{bmatrix} x + \begin{bmatrix} 0.7873 & 0 \\ 0.0031 & 0 \\ 0 & 2.026 \end{bmatrix} u,$$

$$y = \begin{bmatrix} 1 & 0 & 0 \\ 0 & 1 & 0 \end{bmatrix} x,$$

(3.74)

where the state, input and output are deviation variables: $x = [\delta C_a, \delta T, \delta T_j]^T$, $u = [\delta F, \delta F_j]^T$, $y = [\delta C_a, \delta T]^T$. The above linearised model at the nominal operating point, denoted as $G(s)$, is used to design a linear robust controller for the nonlinear plant. If there are unmeasured variations of concentration C_{a0}, temperature T_0 of the inlet flow and the holdup of the tank $V(t)$, the process will be operating at different (perturbed) operating points. Assume that the linearised models $G_t(s)$ at the perturbed operating points belong to a set Π. Then the multiplicative uncertainty caused by variations in the operating point $\Delta_L(s) = [G_t(s) - G(s)]\,G(s)^{-1}$ can be estimated. The controller is required to satisfy the following specifications:

Robust stability: The closed-loop system should be stable for

- Uncertainty $\Delta_L(s)$ caused by variations in the operating point, $C_{a0}(t)$, $T_0(t)$ and $V(t)$: they may vary by $\pm 4\%$, $\pm 0.5\%$, $\pm 1\%$, respectively.
- Unstructured high-order uncertainty arising from the ignored high-order dynamics of the process and other uncertainty factors, the gain bound of this part of uncertainty is estimated as

$$\bar{\sigma}\left(\Delta_N(j\omega)\right) \leq \left| \frac{10j\omega + 1}{j\omega + 1000} \right|, \quad \forall \ \omega.$$

(3.75)

Varying C_{a0}, T_0 and V simultaneously, the frequency response data of the linearised models $G_t(j\omega)$ are obtained, from which the regions of frequency response of $\Delta_L(j\omega)$ are calculated (similar to Example 3.2). The gain bound of uncertainty is then estimated at different frequencies. It is found that $\Delta_L(s)$ has quite a large gain at low frequencies and is approximately 170% at steady state. By curve fitting, the gain bound of total uncertainty $\Delta(s) = \Delta_L(s) + \Delta_N(s)$ is approximated by the following first-order model:

$$\bar{\sigma}[\Delta(j\omega)] \leq \left| \frac{j\omega + 34}{0.05j\omega + 20} \right| \quad \forall \ \omega.$$

(3.76)

Choose the upper sector bound as $b = 100$. The sector-based passivity index can be estimated from the frequency response regions of $\Delta_L(j\omega)$ by using Theorem 3.9. It is found that at steady state, $\nu_{S-}\left(\Delta_L(s), 100, 0\right) = 0.84$, which is significantly smaller than the gain bound of 1.7. From Property 3.10,

$$\nu_{S-}(\Delta_L + \Delta_N, \omega, 150) \le \nu_{S-}(\Delta_L(s), \omega, 100) + \nu_{S-}(\Delta_N(s), \omega, 50)$$
$$\le \nu_{S-}(\Delta_L(s), \omega, 100) + \bar{\sigma}(\Delta_N(j\omega))$$
$$< \text{Re}\left(\frac{j\omega + 11}{0.05 j\omega + 13}\right), \quad \forall \omega. \tag{3.77}$$

Nominal performance: The concentration C_a and temperature T are required to be controlled at 0.252 lb·mol /ft^3 and 597.7 °R, respectively. In addition

- The steady-state offset is less than $1/1000$ of the disturbance value.
- Disturbances should be rejected within 6 hours.

Control design: To achieve the nominal performance specification, we can choose

$$W_1(s) = \begin{bmatrix} \frac{s+1000}{1000s+1} & 0 \\ 0 & \frac{s+1000}{1000s+1} \end{bmatrix}. \tag{3.78}$$

Weighting function $W_2(s)$ reflects the passivity bound of the uncertainty and should satisfy

$$-\nu_{S-}(W_2(s), \omega, b_w) > \nu_{S-}(\Delta(s), \omega, 150). \tag{3.79}$$

According to (3.77), with $w_b = 150$,

$$W_2(s) = \begin{bmatrix} \frac{s+8.5}{0.05s+10} & 0 \\ 0 & \frac{s+8.5}{0.05s+10} \end{bmatrix}. \tag{3.80}$$

Other design parameters can be chosen rather arbitrarily. Here we use $w_3 = 0.01$, and $w_4 = 1.1$.

The controller is then obtained following Procedure 3.16. The controller parameters can be downloaded from companion website for the book at springer.com. A conventional $\mathcal{H}_\infty$ controller design for this system cannot achieve the required nominal performance because of the large uncertainty at low frequencies. The $\mathcal{H}_\infty$ controller was designed such that the system was robustly stable with the best possible nominal performance. The following mixed sensitivity problem is solved:

$$\left\| \begin{bmatrix} W_{H1}S & W_{H2}T \end{bmatrix}^T \right\|_\infty < 1, \tag{3.81}$$

where

$$W_{H1}(s) = \begin{bmatrix} \frac{5s+50}{100s+43} & 0 \\ 0 & \frac{5s+50}{100s+43} \end{bmatrix}, \quad W_{H2}(s) = \begin{bmatrix} \frac{s+360}{200} & 0 \\ 0 & \frac{s+360}{200} \end{bmatrix}.$$

Simulation results: Closed-loop simulation using the nonlinear model and two different controllers has been performed. Assume that at the end of the first hour, C_{a0} changes to 0.48 lb · mol/ft^3, T_0 changes to 603 °R

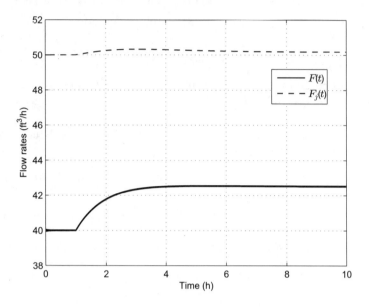

Fig. 3.12. Outputs of the passivity-based controller

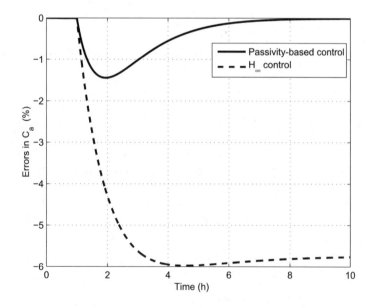

Fig. 3.13. Control errors in $C_a(t)$

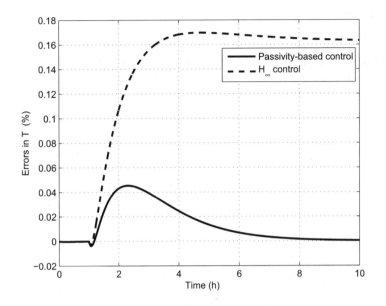

Fig. 3.14. Control errors in $T(t)$

and V changes to 51 ft^3, simultaneously. Figure 3.12 shows the outputs of the passivity-based controller. Figures 3.13 and 3.14 show the normalised control errors (in percentage) of the $C_a(t)$ and $T(t)$ produced by the two control methods. Compared with the control errors of conventional $\mathcal{H}_\infty$ control, the benefits of using the method developed in this section can be seen: The ITAE of the outlet controlled variables for the first 10 hours is 0.0545 for the passivity-based approach, compared to 2.916 for $\mathcal{H}_\infty$ control. The large gain of multiplicative uncertainty at steady state prohibits the use of controllers with high steady-state gain in the $\mathcal{H}_\infty$ control design, leading to a large offset.

3.4 Combining Passivity with the Small Gain Condition

From the discussion in Section 3.3.2, we can see that the gain constraint on the system M (the system "seen" by the uncertainty) at low frequencies affects the achievable control performance because high controller gain is required only at low frequencies to achieve desirable dynamic control performance. Intuitively, we can develop a robust control strategy which satisfies the passivity condition at low frequencies and the small gain condition at high frequencies. This leads to a simple approach which is less conservative than the robust control approaches based on the small gain condition alone. This section is based mainly on [9].

First, we need to define positive realness and the passivity-based uncertainty measure over a frequency band:

Definition 3.18 (Positive realness over a frequency band [9]). *A stable linear system with a transfer function $T(s)$ is said to be strictly positive real over a frequency band $[\omega_1, \omega_2]$ if*

$$T'(j\omega) + T'^*(j\omega) > 0, \quad \forall\, \omega \in [\omega_1, \omega_2]. \tag{3.82}$$

Definition 3.19. *For a given stable linear system with a transfer function $T(s)$, its passivity-based uncertainty measure over a certain frequency band $[\omega_1, \omega_2]$ is defined as*

$$\nu_{FB-}\left(T(s), \omega_1, \omega_2\right) \triangleq -\min\left\{ \min_{\omega \in [\omega_1, \omega_2]} \lambda\left(\frac{1}{2}[T(j\omega) + T^*(j\omega)]\right), 0 \right\}. \tag{3.83}$$

Obviously, system $T'(s) = T(s) + \nu_{FB-}(T(s), \omega_1, \omega_2)I$ is stable and strictly positive real over the frequency band $[\omega_1, \omega_2]$. If the ν_{FB-} index of an uncertainty system is small, the uncertainty system is said to be *near passive*. In particular, for multiplicative uncertainties, we have the following definition:

Definition 3.20 ([9]). *A multiplicative uncertainty system $\Delta_M(s)$ is said to be near passive in a certain frequency band $[\omega_1, \omega_2]$ if $\Delta_M(s)$ is stable and*

$$\nu_{FB-}\left(\Delta_M(s), \omega_1, \omega_2\right) < 1. \tag{3.84}$$

Clearly, if $\underline{\sigma}\left(\Delta_M(j\omega)\right) < 1$ for $\omega \in [\omega_1, \omega_2]$, then, $\Delta_M(s)$ is near passive in the frequency band $[\omega_1, \omega_2]$.

3.4.1 Robust Stability Condition Based on Passivity and Gain

Now let us look at the robust stability condition that combines the passivity and gain condition in different frequency bands.

Theorem 3.21 (Robust stability based on both passivity and system gain [9]). *For the interconnected system shown in Figure 3.3, the closed-loop system is stable if*

1. *$\Delta(s)M(s)$ is strictly proper; and*
2. *$M(s)$ is strictly positive real in frequency band $[0, \omega_1]$, $\Delta(s)$ is stable and positive real in the same frequency band and*

$$\bar{\sigma}[\Delta(j\omega)]\bar{\sigma}[M(j\omega)] < 1 \tag{3.85}$$

over the frequency band (ω_1, ∞).

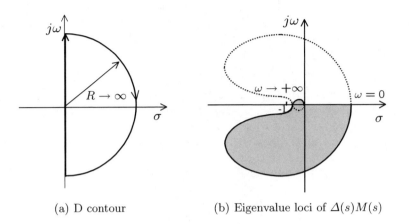

(a) D contour   (b) Eigenvalue loci of $\Delta(s)M(s)$

Fig. 3.15. Nyquist plot

Proof. Since both $M(s)$ and $\Delta(s)$ do not have poles on either the right half plane or the imaginary axis, the generalized Nyquist stability condition reduces to the condition that the characteristic value loci should never encircle the critical point $-1 + 0j$ when s travels on the standard Nyquist D-contour. From Condition 1, the eigenvalues of $\Delta(s)M(s)$ will be at the origin while s traverses the right half circle with the infinite radius of the D-contour, as shown in Figure 3.15a. Thus only the imaginary axis part of the D-contour needs to be considered, and the stability of the closed-loop system can be determined by proving the following two statements:

1. The eigenvalue loci of $\Delta(j\omega)M(j\omega)$ for $\omega \in [0, \omega_1]$ do not encircle the critical point.
2. The eigenvalue loci of $\Delta(j\omega)M(j\omega)$ for $\omega \in (\omega_1, \infty)$ do not encircle the critical point either.

Proof of Statement 1: Since $\Delta(s)$ is stable and passive, and $M(s)$ is strictly positive real in $[0, \omega_1]$,

$$M(j\omega) + M^*(j\omega) > 0 \quad \text{and} \quad \Delta(j\omega) + \Delta^*(j\omega) \geq 0, \quad \forall \, \omega \in [0, \omega_1]. \quad (3.86)$$

Consequently, the following two inequalities hold:

$$\text{Re}\,(x^* M(j\omega)x) > 0 \text{ and } \text{Re}\,(x^* \Delta^*(j\omega)x) \geq 0, \quad (3.87)$$

for all $x \neq 0$ and all real $\omega \in [0, \omega_1]$. For any real eigenvalue λ of $\Delta(j\omega)M(j\omega)$,

$$\lambda x = \Delta(j\omega)M(j\omega)x \quad (3.88)$$

with $x \neq 0$. It can be shown that

$$\text{Re}\,(\lambda x^* M^*(j\omega)x) = \text{Re}\,(x^* M(j\omega)^* \Delta(j\omega)M(j\omega)x) = \text{Re}\,(y^* \Delta(j\omega)y) \geq 0,$$
$$(3.89)$$

where $y = M(j\omega)x \neq 0$, since $M(j\omega)$ is not singular. This leads to the relationship:

$$\text{Re}\,(\lambda x^* M(j\omega)^* x) = \lambda\,\text{Re}\,(x^* M(j\omega)^* x) \geq 0. \qquad (3.90)$$

Because λ is real and $\text{Re}\,(x^* M(j\omega)^* x) > 0$, $\lambda \geq 0$, that is, the eigenvalue of $\Delta(j\omega)M(j\omega)$ does not cross the negative real axis, and thus does not encircle the critical point $-1 + j0$ while s traverses the imaginary axis from $j0$ to $j\omega_1$. *Proof of Statement 2*: Since

$$\begin{aligned}\left|\bar{\lambda}\left(\Delta(j\omega)M(j\omega)\right)\right| &\leq \bar{\sigma}\left(\Delta(j\omega)M(j\omega)\right) \\ &\leq \bar{\sigma}\left(\Delta(j\omega)\right)\bar{\sigma}\left(M(j\omega)\right) < 1,\end{aligned} \qquad (3.91)$$

the moduli of the eigenvalues of $\Delta(j\omega)M(j\omega)$ are always less than 1 for any $\omega \in (\omega_1, \infty)$.

By combining the proofs for statements (1) and (2), it is concluded that the eigenvalue loci of $\Delta(j\omega)M(j\omega)$ will never encircle the critical point $(-1, j0)$ for all $\omega \in [0, \infty)$, and thus the closed-loop system of Δ and M is stable.

From the above discussion, it can be seen that the small gain condition prevents the eigenvalue loci of $\Delta(j\omega)M(j\omega)$ from encircling the critical point by limiting the moduli of the eigenvalues of $\Delta(j\omega)M(j\omega)$ while the positive-real condition limits the arguments of the eigenvalues to the open interval of $(-\pi, \pi)$. Typical eigenvalue loci that satisfy Theorem 3.21 are shown in Figure 3.15b.

In robust control design, it is the low and midfrequencies at which high controller gain is needed to achieve good dynamic performance and the competition between robust stability and control performance occurs. Therefore, Theorem 3.21 provides a simple yet powerful approach to high-performance robust control, particularly when the uncertainties have significant gain at low frequencies, such as uncertainties caused by parameter variations exemplified by the CSTR problem in Section 3.3.5. If the uncertainty $\Delta(s)$ is near passive at low frequencies, then we can have the following proposition, which can be obtained using loop shifting:

Proposition 3.22. *For the interconnected system shown in Figure 3.2, the closed-loop system is stable if $M(s)$ is strictly proper and $\Delta_M(s)$ is stable with $\nu_{FB-}(\Delta_M(s), 0, \omega_1) \leq \nu_{max} < 1$,*

$$\tilde{M}(s) = M(s)\left[I - \nu_{max}M(s)\right]^{-1} \qquad (3.92)$$

is strictly positive real in frequency band $[0, \omega_1]$ and

$$\left[\bar{\sigma}(\Delta(j\omega)) + \nu_{max}\right]\bar{\sigma}[\tilde{M}(j\omega)] < 1 \qquad (3.93)$$

in frequency band (ω_1, ∞).

Theorem 3.21 appears similar to the Small-phase-small-gain Theorem given in [96]. The latter combines the small phase and small gain conditions in different frequency bands. Define the phase modifier

$$\psi_m(\omega) \triangleq \tan^{-1}\left[\frac{[c_T(\omega) - 1]c_\Delta(\omega)}{1 - [c_T(\omega) - 1]c_\Delta(\omega)}\right], \tag{3.94}$$

where $c_T(\omega)$ and $c_\Delta(\omega)$ are condition numbers of $M(j\omega)$ and $\Delta(j\omega)$. Adopting the phase definition given in Definition 2.26, the Small-phase-small-gain Theorem can be briefly stated as follows: if both $\Delta(s)$ and $M(s)$ are stable, the sum of the principal phase of $\Delta(s)$ and $M(s)$ is in the open interval of $(-\pi + \psi, \pi - \psi)$ in the low-frequency band and the small gain condition is ensured in the high-frequency band, then the closed-loop system is stable.

Theorem 3.21 is different from the small-phase-small gain condition because it is not based on the actual phase of the subsystems but on the phase interval (positive-real condition). While the passivity condition is less flexible, it is more applicable to control design since it is easier to design a controller to render a MIMO system positive real rather than to minimize its phases. Another problem of the Small-phase-small-gain Theorem is that the small phase condition for MIMO systems can be conservative if the condition number of either $\Delta(s)$ or $M(s)$ is large, leading to a large phase modifier. For example, if the condition number of $T(j\omega_1)$ is 1.05 and the condition number of $\Delta(j\omega_1)$ is 10, then the phase modifier will be approximately $\pi/2$. Therefore, the small phase condition is equivalent to the sum of the phase of $M(j\omega_1)$ and $\Delta(j\omega_1)$ in the interval $(-\pi/2, \pi/2)$, which is very restrictive.

3.4.2 Control Synthesis

Problem formulation

Consider a nominal process model $G(s) := (A_g, B_g, C_g, D_g)$ with multiplicative uncertainty $\Delta_M(s)$, as shown in Figure 3.2. The uncertainty is assumed to be stable and near-passive at least in the low-frequency range. A robust controller $K(s)$ is to be designed to stabilize the plant and achieve a certain level of nominal performance. More specifically, the closed-loop system should be robustly stable under the uncertainty Δ bounded by the following conditions:

$$\nu_{FB-}(\Delta(s), 0, \omega_1) = \nu_{\max} < 1, \tag{3.95}$$

$$\bar{\sigma}[\Delta(j\omega)] < |w_\Delta(j\omega)|, \quad \forall \omega \in [\omega_1, +\infty]. \tag{3.96}$$

Passivating control design

Because multiplicative uncertainty is assumed, the system seen by the uncertainty $M(s)$ is the complementary sensitivity function $T(s)$. If the uncertainty is positive real at low frequencies, the controller should render the system

$T(s)$ strictly positive real at these frequencies to achieve robust stability. For systems with near passive uncertainties whose $\nu_-(\Delta(s)) < \nu_{\max} < 1$ in a certain low-frequency band, the controller $K(s)$ should render the system $\tilde{T} = T(T - \nu_{\max}I)^{-1}$ strictly positive real in the same frequency band. Similar to the approach in Section 3.3.4, this problem can be converted into an $\mathcal{H}_\infty$ problem by using the *Cayley transformation* given in (3.63):

$$T'(s) = \left[\zeta I - \tilde{T}(s)\right]\left[\zeta I + \tilde{T}(s)\right]^{-1}. \tag{3.97}$$

Here we choose $\zeta = (1 - \nu_{\max})^{-1}$ to obtain a direct relationship between the nominal performance and positive realness and thus to simplify the controller design procedure. A weighting function $W_1(s)$ can be applied to T' after the Cayley transformation such that the positive realness is achieved in a given frequency band. It is chosen as a diagonal transfer function, $W_1(s) = w_1(s)I$, where $w_1(s)$ is a stable, proper and minimum phase transfer function with $|w_1(j\omega)| > 1$ for the frequency band over which $\tilde{T}(s)$ is strictly positive real. Therefore, the controller $K(s)$ which satisfies the $\mathcal{H}_\infty$-norm condition,

$$\|W_1 T'\|_\infty < 1, \tag{3.98}$$

will achieve strictly positive realness of $\tilde{T}(s)$ in the frequency band.

The positive real condition itself permits controllers with very large gains. The weighting function $W_2(s)$ is then used to limit the controller gain when the controller is connected to the process. $W_2(s)$ must be stable, proper, minimum phase and usually diagonal. The inequality,

$$\left\|2(1 - \nu_{\max}) W_2 K \left[I + 2(1 - \nu_{\max}) GK\right]^{-1}\right\|_\infty < 1, \tag{3.99}$$

is satisfied in the controller design.

Assume that

$$W_1(s) := (A_{w1}, B_{w1}, C_{w1}, D_{w1})$$

and

$$W_2(s) := (A_{w2}, B_{w2}, C_{w2}, D_{w2}).$$

Designing a controller that satisfies both (3.98) and (3.99) is equivalent to the following $\mathcal{H}_\infty$ control problem:

Problem 3.23. Find a stabilizing $K(s)$ for the following augmented plant P:

$$\dot{x} = \begin{bmatrix} A_g & 0 & 0 \\ B_{w1}C_g & A_{w1} & 0 \\ 0 & 0 & A_{w2} \end{bmatrix} x + \begin{bmatrix} 0 \\ B_{w1} \\ 0 \end{bmatrix} w + \begin{bmatrix} B_g \\ B_{w1}D_g \\ B_{w2} \end{bmatrix} u,$$

$$z = \begin{bmatrix} D_{w1}C_g & C_{w1} & 0 \\ 0 & 0 & C_{w2} \end{bmatrix} x + \begin{bmatrix} D_{w1} \\ 0 \end{bmatrix} w + \begin{bmatrix} D_{w1}D_g \\ D_{w2} \end{bmatrix} u,$$

$$y = \begin{bmatrix} 2(1 - \nu_{\max})C_g & 0 & 0 \end{bmatrix} x - 2(1 - \nu_{\max})Iw + 2(1 - \nu_{\max})D_g u,$$

$$\tag{3.100}$$

such that the closed-loop system is internally stable and $\|T_{zw}\|_\infty < 1$, where $T_{zw}(s)$ denotes the system from w to z.

Tools such as the MATLAB® *Robust Control Toolbox* can be used to solve this problem.

Nominal performance

Although robust stability is based on passivity, the performance can be measured using the maximum singular values of the sensitivity function $S(s)$. In the previous section, we have limited the maximum singular values of $T'(j\omega)$ to a certain frequency band to achieve positive realness. The following will show that the maximum singular values (and thus the ∞-norm) of the sensitivity function is also bounded by that of $T'(j\omega)$.

For nonsingular T',

$$
\begin{aligned}
T' &= \left(\frac{1}{1-\nu_{\max}}I - \tilde{T}\right)\left(\frac{1}{1-\nu_{\max}}I + \tilde{T}\right)^{-1} \\
&= \left\{\frac{1}{1-\nu_{\max}}I - GK\left[I + (1-\nu_{\max})GK\right]^{-1}\right\} \\
&\quad \cdot \left\{\frac{1}{1-\nu_{\max}}I + GK\left[I + (1-\nu_{\max})GK\right]^{-1}\right\}^{-1} \\
&= \left\{I - (1-\nu_{\max})GK\left[I + (1-\nu_{\max})GK\right]^{-1}\right\} \\
&\quad \cdot \left\{I + (1-\nu_{\max})GK\left[I + (1-\nu_{\max})GK\right]^{-1}\right\}^{-1} \\
&= \left\{I - (1-\nu_{\max})GK\left[I + (1-\nu_{\max})GK\right]^{-1}\right\}\left[I + (1-\nu_{\max})GK\right] \\
&\quad \cdot \left[I + (1-\nu_{\max})GK\right]^{-1}\left\{I + (1-\nu_{\max})GK\left[I + (1-\nu_{\max})GK\right]^{-1}\right\}^{-1} \\
&= \left[I + 2(1-\nu_{\max})GK\right]^{-1}.
\end{aligned}
$$

$$(3.101)$$

Therefore,

$$
\begin{aligned}
\bar{\sigma}\left(S(j\omega)\right) &= \bar{\sigma}\left(I + G(j\omega)K(j\omega)\right)^{-1} \\
&\leq \frac{2(1-\nu_{\max})\bar{\sigma}\left(T'(j\omega)\right)}{1 - |1 - 2\nu_{\max}|\,\bar{\sigma}\left(T'(j\omega)\right)}.
\end{aligned}
$$

$$(3.102)$$

Since $0 \leq \nu_{\max} < 1$ and $\bar{\sigma}\left(T'(j\omega)\right) < 1$,

$$
|1 - 2\nu_{\max}|\,\bar{\sigma}\left(T'(j\omega)\right) < 1. \tag{3.103}
$$

It can be concluded that the nominal performance is bounded by $\bar{\sigma}\left(T'(j\omega)\right)$ in the frequency range where strict positive realness is achieved. The only cost is high controller gain. By choosing an appropriate weighting function $W_1(s) = w_1(s) I$ for (3.100), we can obtain specified performance as well as the positive realness of $\tilde{T}$. If $\bar{\sigma}(S(j\omega)) < \alpha$ is required in some frequency range, $w_1(s)$ should satisfy the following inequality in that frequency range:

$$|w_1(j\omega)| > \frac{2(1 - \nu_{\max})}{\alpha} - 2\nu_{\max} + 1. \tag{3.104}$$

In general, $w_1(s)$ could be chosen such that

1. $w_1(s)$ has high gain at low frequencies where good performance is required. Especially, $|w_1(0)| \to \infty$ to ensure small steady-state error.
2. $w_1(s)$ has enough frequency bandwidth for fast command response and/or disturbance rejection.
3. $|w_1(j\omega)| > 1$ for the frequency band in which passivity should be achieved.
4. $|w_1(j\omega)|$ is small at high frequencies.

Combining with the small gain condition

At high frequencies where the uncertainty may not be passive, the small gain condition

$$\bar{\sigma}\left([w_\Delta(j\omega) + \nu_{\max}]\tilde{T}(j\omega)\right) < 1 \tag{3.105}$$

applies. However, it is difficult to solve the following $\mathcal{H}_\infty$ problem for two different system $\tilde{T}(s)$ and $T'(s)$:

$$\left\|\begin{array}{c} W_1 T' \\ (w_\Delta + \nu_{\max})\tilde{T} \end{array}\right\|_\infty < 1. \tag{3.106}$$

The approach used here is to limit the gain of the controller at high frequencies. This can be seen from the following inequality:

$$\bar{\sigma}\left(\tilde{T}(s)\right) = \bar{\sigma}\left\{G(s)K(s)\left[I + (1 - \nu_{\max})G(s)K(s)\right]^{-1}\right\}$$
$$\approx \bar{\sigma}\left(G(s)K(s)\right) \le \bar{\sigma}\left(G(s)\right)\bar{\sigma}\left(K(s)\right) \tag{3.107}$$

for $\bar{\sigma}(GK) \ll 1$ and $\nu_{\max} < 1$.

Because the passivating controller $K(s)$ stabilizes the plant $G(s)$, robust stability at high frequencies can be obtained by using an appropriate weighting function $W_2(s)$ in (3.100). $W_2(s)$ should have small gain at low frequencies and large gain at high frequencies.

Control design procedure

The controller can be designed by following the step-by-step design procedure:

Procedure 3.24 (Robust control design [9])

1. *Determine $W_1(s)$ according to (3.104) and choose $W_2(s)$ such that its gain is small at low frequencies and large at high frequencies.*
2. *Solve the $\mathcal{H}_\infty$ problem of the augmented plant in (3.100) to obtain the controller $K(s)$.*
3. *If the gain of $W_2(s)$ is too high such that no controller can be found to meet the passivity and performance requirements, decrease the gain of $W_2(s)$, and go to step (2) to redesign $K(s)$.*
4. *Check whether $\tilde{T}(s)$ meets the requirements of robust stability at high frequencies. If not, increase the gain of $W_2(s)$ at high frequencies and go to step (2) to redesign $K(s)$.*
5. *If both the passivity/performance condition at low frequencies and the small gain condition at high frequencies are satisfied, the final robust controller $K(s)$ is obtained.*

Some trial and error will be required to find the appropriate weighting functions $W_1(s)$ and $W_2(s)$. They should be stable and minimum phase and can have the following structures:

$$
W(s) = w(s)I = \frac{k \prod_{i=1}^{m} (\tau_{Zi} s + 1)}{\prod_{j=1}^{n} (\tau_{Pj} s + 1)} I. \tag{3.108}
$$

The order of the weighting functions is generally less than 3 to avoid high-order controllers. The gain of the weighting functions can be shaped by adding a pole $-1/\tau_{Pj}$, adding a zero $-1/\tau_{Zi}$ and tuning k to shift the gain plot upward and downward.

3.4.3 Robust Control of a Mixing System

The control of a mixing system is used to illustrate the proposed control design method [9]. As shown in Figure 3.16, the tank is fed with two inlet flows with flow rates $F_1(t)$ and $F_2(t)$. Both inlet flows contain one dissolved material with concentrations c_1 and c_2, respectively. The outlet flow rate is $F(t)$. Assume that the tank is well stirred so that the concentration of the outlet flow is the same as the concentration in the tank. The inlet flow rates $F_1(t)$ and $F_2(t)$ are manipulated to control both the flow rate $F(t)$ and the outlet concentration $c(t)$ at desired values.

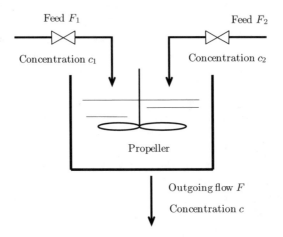

Fig. 3.16. Mixing system

Process model

The linearised model at the point $(V = V_0,\ c = c_0,\ F_1 = F_{10},\ F_2 = F_{20},$ $c_1 = c_{10},\ c_2 = c_{20})$ is as follows [73]:

$$\dot{x} = \begin{bmatrix} -\frac{1}{2\theta} & 0 \\ \frac{(F_{10}+F_{20})c_0-(c_{10}F_{10}+c_{20}F_{20})}{V_0^2} & \frac{F_{10}+F_{20}}{V_0} \end{bmatrix} x + \begin{bmatrix} 1 & 1 \\ \frac{c_{10}-c_0}{V_0} & \frac{c_{20}-c_0}{V_0} \end{bmatrix} u,$$

$$y = \begin{bmatrix} \frac{1}{2\theta} & 0 \\ 0 & 1 \end{bmatrix} x,$$

$$(3.109)$$

where $x = \begin{bmatrix} \delta V, & \delta c \end{bmatrix}^T,\ y = \begin{bmatrix} \delta F, & \delta c \end{bmatrix}^T,\ u = \begin{bmatrix} \delta F_1, & \delta F_2 \end{bmatrix}^T$ and $\theta = \frac{\sqrt{V_0}S}{k}$. The plant model linearised at a nominal point $(F_1(0) = 1.5 \text{ m}^3/\text{min},\ F_2(0) = 2.0 \text{ m}^3/\text{min},\ F(0) = 3.5 \text{ m}^3/\text{min},\ c_1(0) = 1 \text{ kmol/m}^3,\ c_2(0) = 2 \text{ kmol/m}^3,\ c(0) = 1.57 \text{ kmol/m}^3,\ V_0/F_0 = 1 \text{ min})$ is

$$\dot{x} = \begin{bmatrix} -0.01 & 0 \\ 0 & -0.02 \end{bmatrix} x + \begin{bmatrix} 1 & 1 \\ -0.25 & 0.75 \end{bmatrix} u,$$

$$y = \begin{bmatrix} 0.01 & 0 \\ 0 & 1 \end{bmatrix} x.$$

$$(3.110)$$

Denote $G(s)$ as the transfer function of the above nominal system. The uncertain plant $G_t(s) \in \Pi$ due to the uncertainties in c_1 and c_2. The gain of the linear uncertainty $\Delta_L(s) = [G_t(s) - G(s)]G(s)^{-1}$ can be large at low frequencies when c_1 and c_2 change significantly. At the nominal operating point, c_1 is less than and c_2 is greater than the concentration c in the outlet stream.

It is found that if c_1 changes in the interval $[0, 1]$ and c_2 changes in the interval $[2, +\infty)$ simultaneously, then the uncertainty $\Delta_L(s)$ is passive at any frequency. If $0 < c_1 < 1.57$ and $1.57 < c_2 < 2$, then the linear uncertainty is near-passive.

Design specifications

It is required to control the product concentration and the flow rate at the nominal operating point to meet the following specifications:

Robust stability: The closed-loop system should be stable for

1. The uncertainty $\Delta_L(s)$ caused by variations in both inlet concentrations: $c_1(t) \in [0.5, 1.2]$ kmol/m^3 and $c_2(t) \in [1.7, 3.0]$ kmol/m^3.
2. Unstructured uncertainty $\Delta_N(s)$ arising from neglected high-order dynamics. The overall multiplicative uncertainty $\Delta(s)$ is the total effect of $\Delta_L(s)$ and $\Delta_N(s)$ and is found to be bounded by

$$\bar{\sigma}\left[\Delta(j\omega)\right] \leq \left|\frac{j\omega + 22.5}{15}\right|. \tag{3.111}$$

While the uncertainty has large gain at low frequencies (150% at steady state), it is near-passive in the low-frequency band:

$$\nu_{FB-}\left(\Delta(s), 0, 3\right) = 0.6. \tag{3.112}$$

Performance:

1. The steady-state offset is less than $1/1000$ of the disturbance value.
2. Disturbances should be rejected within 5 minutes.

Control design

Following the design procedure given in Section 3.4.2, the weighting functions $W_1(s)$ and $W_2(s)$ are chosen as follows to meet the performance and robust specifications:

$$W_1(s) = \begin{bmatrix} \frac{1000(s+2)}{(s+0.001)(s+30)^2} & 0 \\ 0 & \frac{1000(s+2)}{(s+0.001)(s+30)^2} \end{bmatrix}, \tag{3.113}$$

$$W_2(s) = \begin{bmatrix} \frac{100s+1}{s+200000} & 0 \\ 0 & \frac{100s+1}{s+200000} \end{bmatrix}. \tag{3.114}$$

The controller is obtained by solving Problem 3.23. The controller parameters can be downloaded from the companion website for the book. The singular value plot of the controller is shown in Figure 3.17. The frequency-dependent IFP index of system $T(s)$ ("seen" by the uncertainty) is shown in Figure 3.18,

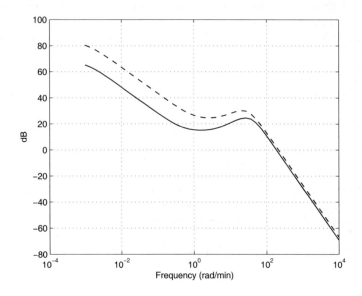

Fig. 3.17. Singular values of the controller [9]

which shows that system $T(s)$ has excessive IFP at low frequencies up to 20 rad/min and is nonpassive at higher frequencies where the small gain condition applies. The singular value plot of the sensitivity function, which indicates the nominal performance, is shown in Figure 3.19, which shows that the bandwidth of the sensitivity function is similar to the frequency band over which the excessive IFP of $T(s)$ is achieved.

Simulation results

Closed-loop simulation using the nonlinear model and the passivity-based controller has been performed. Assume that at the 5th minute, c_1 decreases from 1 kmol/m^3 to 0.5 kmol/m and c_2 increases from 2 kmol/m^3 to 3 kmol/m^3. The controller outputs are shown in Figure 3.20. The control errors in the outlet concentration are shown in Figure 3.21.

An $\mathcal{H}_\infty$ controller cannot achieve the required performance specification because of the large uncertainty at low frequencies. If the allowed variations of c_1 and c_2 are limited to a smaller range, e.g., $0.6 < c_1 < 1.2$ and $1.7 < c_2 < 2.6$, then the required performance can be achieved by an $\mathcal{H}_\infty$ controller.

3.5 Passive Controller Design

Certain classes of process systems, including some mechanical systems (e.g., [126, 136]) and certain thermodynamic systems (e.g., [144]) are inherently

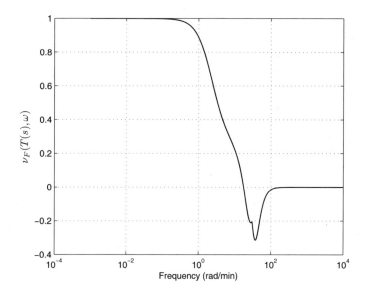

Fig. 3.18. IFP index of $T(s)$

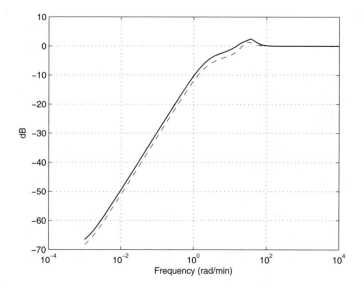

Fig. 3.19. Singular value plot of the sensitivity function

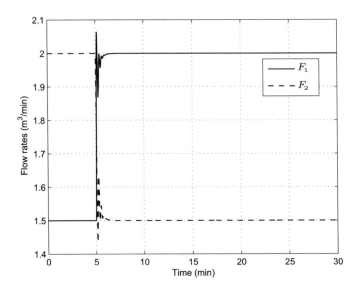

Fig. 3.20. Controller outputs $F_1(t)$ and $F_2(t)$

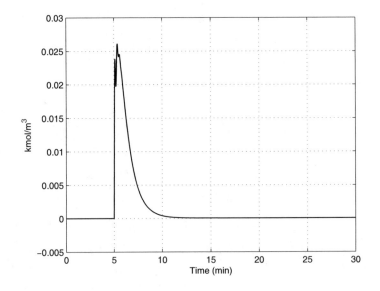

Fig. 3.21. Control errors in $C(t)$

passive. They can be robustly stabilized by strictly passive (or strictly positive real) linear controllers.

Based on [148] and [150], this section presents an approach to synthesis of strictly positive real (SPR) controllers that achieve $\mathcal{H}_\infty$ performance. As we have learned in previous sections, both the SPR and $\mathcal{H}_\infty$ conditions can be written as LMIs by using the positive-real lemma and the bounded-real lemma. Either of the two conditions is a convex problem and can be solved using SDP techniques. However, to solve both conditions simultaneously is a very difficult task. Existing multi-objective design methods (e.g., [41, 62]) do not apply to the SPR/$\mathcal{H}_\infty$ control problem because the SPR and $\mathcal{H}_\infty$ conditions are formulated for different systems (the $\mathcal{H}_\infty$ condition for the closed-loop system and the SPR condition for the controller only).

There are few SPR/$\mathcal{H}_\infty$ control synthesis methods available. The ad hoc approaches for SISO systems were developed by Chen and Wen [29]. Geromel and Gapski proposed to incorporate $\mathcal{H}_\infty$ constraints into an existing SPR/$\mathcal{H}_2$ method [42]. This approach assumes an LQG controller structure and requires simplifying assumptions on the generalized plant models (e.g., no correlation between measurement noise and disturbances and no cross product terms relating the control actions and states in the cost function, etc.). Many generalized plants arising from $\mathcal{H}_\infty$ performance specifications do not satisfy these assumptions, and therefore, these problems cannot be solved using the above approach.

In this section, we present a multivariable SPR/$\mathcal{H}_\infty$ control design approach, which does not require the assumptions in [42]. Only process stabilizability and detectability are required. This approach was first presented in [150] and was detailed in [148].

3.5.1 Problem Formulation

The SPR/$\mathcal{H}_\infty$ control problem can be formulated as follows: Given an LTI generalized plant model:

$$P(s) : \begin{cases} \dot{x} = Ax + B_1 w + B_2 u \\ z = C_1 x + D_{11} w + D_{12} u \\ y = C_2 x + D_{21} w + D_{22} u, \end{cases} \tag{3.115}$$

where $x \in \mathbb{R}^n$, $y \in \mathbb{R}^m$, $u \in \mathbb{R}^m$, $z \in \mathbb{R}^p$, $w \in \mathbb{R}^q$, find an SPR controller $K(s) := (A_k, B_k, C_k, D_k)$ such that the $\mathcal{H}_\infty$ norm of the closed-loop system from w to z is bounded, i.e.,

$$\|T_{zw}\|_\infty < 1. \tag{3.116}$$

Similar to the approach presented in Section 3.3.4, we find all-solution controllers satisfying the $\mathcal{H}_\infty$-performance specification which are parameterized by a free stable contraction map:

$$K(s) = \mathcal{F}_l\left(F(s), U(s)\right), \tag{3.117}$$

where $F(s)$ is the all-solution controller parametrization and $U(s)$ is a contraction map ($\|U\|_\infty < 1$). The SPR/$\mathcal{H}_\infty$ controller can be obtained by finding a stable contraction map $U(s)$ such that $K(s) = \mathcal{F}_l\left(F(s), U(s)\right)$ is SPR. This involves solving a BMI problem with the passive controller constraints represented using the positive-real lemma (Lemma 2.16).

3.5.2 Contraction Map

The dynamics of the contraction map $U(s)$ are absorbed into the final controller $K(s)$. Therefore, to reduce the controller order to the minimum level, a constant contraction map $U(s) = Q$, $Q \in \mathbb{R}^{m \times m}$ is assumed. Suppose that $F(s)$ has the following two-port form:

$$F(s) := \left[\begin{array}{c|c|c} A_F & B_{1F} & B_{2F} \\ \hline C_{1F} & D_{11F} & D_{12F} \\ \hline C_{2F} & D_{21F} & D_{22F} \end{array}\right]. \tag{3.118}$$

Then, the controller $K(s)$ has the following state-space representation:

$$\begin{aligned} A_k &= A_F + B_{2F}QC_{2F}, \\ B_k &= B_{1F} + B_{2F}QD_{21F}, \\ C_k &= C_{1F} + D_{12F}QC_{2F}, \text{ and} \\ D_k &= D_{11F} + D_{12F}QD_{21F}. \end{aligned} \tag{3.119}$$

The constraint on the contraction map $U(s)$ ($\|U(s)\|_\infty < 1$) can be written in the following LMI:

$$\begin{bmatrix} I & Q^T \\ Q & I \end{bmatrix} > 0. \tag{3.120}$$

3.5.3 Synthesis of SPR/$\mathcal{H}_\infty$ Control

From the positive-real lemma, the controller $K(s) := (A_k, B_k, C_k, D_k)$ is SPR if the following matrix inequality holds:

$$\begin{bmatrix} A_k^T P + P A_k & P B_k - C_k^T \\ B_k^T P - C_k & -D_k - D_k^T \end{bmatrix} < 0. \tag{3.121}$$

Define

$$\Psi_k = \begin{bmatrix} A_F^T P + P A_F & P B_{1F} - C_{1F}^T \\ B_{1F}^T P - C_{1F} & -D_{11F} - D_{11F}^T \end{bmatrix}, \tag{3.122}$$

$$Z = \begin{bmatrix} C_{2F} & D_{21F} \end{bmatrix}, \tag{3.123}$$

$$P_F = \begin{bmatrix} B_{2F}^T & -D_{12F}^T \end{bmatrix}, \text{ and} \tag{3.124}$$

$$P_k = P_F \begin{bmatrix} P & 0 \\ 0 & I \end{bmatrix}. \tag{3.125}$$

From (3.119) and (3.121),

$$\Psi_k + Z^T Q^T P_k + P_k^T Q Z < 0. \tag{3.126}$$

Therefore, the SPR/$\mathcal{H}_\infty$ controller can be obtained by solving (3.120) and (3.126) simultaneously for the constant contraction map Q and the Lyapunov matrix P. This is a BMI problem because (3.126) has bilinear terms in two decision variables, Q and P. A solution to this type of problem is given in Section A.1.

By choosing a constant contraction map, the computational load is significantly reduced because the BMI constraint on $\mathcal{H}_\infty$ performance becomes an LMI. In addition, this treatment leads to SPR/$\mathcal{H}_\infty$ controller designs with the same order as the generalized plant.

3.5.4 Control Design Procedure

The design procedure of SPR/$\mathcal{H}_\infty$ control [148] is as follows:

Procedure 3.25 (SPR/$\mathcal{H}_\infty$ control synthesis)

1. *Obtain the all-solution controller parameterization $F(s)$ for the generalized plant $P(s)$ using a standard $\mathcal{H}_\infty$ solver.*
2. *Solve (3.120) and (3.126) simultaneously for the constant contraction map Q and Lyapunov matrix P.*
3. *Obtain the final controller $K(s) = \mathcal{F}_l(F(s), Q)$.*

The second step is most important in the control design procedure. It requires implementing of the following iterative approach to the BMI problem.

1. Specify n_I as the maximum number of iterations allowed.
2. Start with a central $\mathcal{H}_\infty$ controller, *i.e.*, $Q_0 = 0$, and set counter $i = 1$.
3. Calculate $K_i(s) := (A_k, B_k, C_k, D_k)$ using (3.119).
4. Solve the following generalized eigenvalue problem for α_i, P_i and Y_i:
$$\min_{P_i, Y_i} \{\alpha_i\}$$
subject to

$$\begin{bmatrix} A_k^T P_i + P_i A_k & P_i B_k - C_k^T \\ B_k^T P_i - C_k & -D_k - D_k^T \end{bmatrix} < \alpha_i \begin{bmatrix} P_i & 0 \\ 0 & Y_i \end{bmatrix}, \tag{3.127}$$

$$P_i > 0, \tag{3.128}$$

$$Y_i > 0. \tag{3.129}$$

5. With α_i fixed, solve the following optimization problem for P_i and Y_i:
$$\min_{P_i, Y_i} \{\mathrm{Tr}(P_i) + \mathrm{Tr}(Y_i)\}$$
subject to

$$\begin{bmatrix} A_k^T P_i + P_i A_k - \alpha_i P_i & P_i B_k - C_k^T \\ B_k^T P_i - C_k & - D_k - D_k^T - \alpha_i Y_i \end{bmatrix} < 0, \qquad (3.130)$$

$$P_i > 0, \qquad (3.131)$$

$$Y_i > 0. \qquad (3.132)$$

6. With P_i and Y_i fixed, solve the following optimization problem for α_i and Q_i:

$$\min_{Q_i} \{\alpha_i\}$$

subject to

$$\begin{bmatrix} T_{1,i} & T_{2,i} \\ T_{3,i} & T_{4,i} \end{bmatrix} < 0, \qquad (3.133)$$

$$\begin{bmatrix} I & Q_i^T \\ Q_i & I \end{bmatrix} > 0, \qquad (3.134)$$

where

$$T_{1,i} = A_F^T P_i + P_i A_F + C_{2F}^T Q_i^T B_{2F}^T P_i + P_i B_{2F} Q_i C_{2F} - \alpha_i P_i,$$

$$T_{2,i} = P_i B_{1F} + P_i B_{2F} Q_i D_{21F} - C_{1F}^T - C_{2F}^T Q_i^T D_{12F}^T,$$

$$T_{3,i} = B_{1F}^T P_i + D_{21F}^T Q_i^T B_{2F}^T P_i - C_{1F} - D_{12F} Q_i C_{2F}, \text{ and}$$

$$T_{4,i} = -D_{11F} - D_{11F}^T - D_{12F} Q_i D_{21F} - D_{21F}^T Q_i^T D_{12F}^T - \alpha_i Y_i.$$

7. If $\alpha_i < 0$, the algorithm converges to a feasible solution $Q = Q_i$. The solution is found. Stop.

If $\alpha_i \geq 0$:

If $i \leq n_I$, let $i = i + 1$ and then go to step (c).

Otherwise, the control synthesis problem cannot be efficiently solved using the proposed algorithm.

For all iterative methods tackling BMI problems, global convergence cannot be guaranteed. In the SPR/$\mathcal{H}_\infty$ control problem, a global optimum is often not a major concern provided that a local minimum of $\alpha < 0$ can be found. By alternately minimizing α and $\text{Tr}(P) + \text{Tr}(Y)$, the convergence of the proposed algorithm is expedited. As in any iterative method, the choice of the initial point is important for the effectiveness and efficiency of the iterative algorithm. Our experience indicates that the central $\mathcal{H}_\infty$ controller ($Q = 0$) is a good starting point.

3.5.5 Illustrative Example

We illustrate the above control design approach using the multivariable passive process system given in [127]. This process can be robustly stabilized by an SPR controller. With the $\mathcal{H}_\infty$ performance specification, this process has the following augmented plant:

$$A = A^T = \begin{bmatrix} -1.9092 & -1.4588 & -1.0902 & -1.6758 \\ -1.4588 & -1.9149 & -0.7137 & -1.4493 \\ -1.0902 & -0.7137 & -0.8758 & -0.7530 \\ -1.6758 & -1.4493 & -0.7530 & -1.7289 \end{bmatrix},$$

$$B_1 = C_1^T = \begin{bmatrix} 0.9516 & 0.4010 & 0.0431 & 0.4776 \\ 0.2603 & 0.4866 & 0.3709 & 0.1291 \\ 0.5147 & 0.7505 & 0.6933 & 0.4838 \\ 0.6363 & 0.1262 & 0.9358 & 0.9456 \end{bmatrix},$$

$$(3.135)$$

$$D_{11} = D_{11}^T = \begin{bmatrix} 0.8437 & 0.9280 & 0.7998 & 0.4248 \\ 0.9280 & 0.4022 & 0.1510 & 0.6874 \\ 0.7998 & 0.1510 & 0.0430 & 0.7157 \\ 0.4248 & 0.6874 & 0.7157 & 0.4577 \end{bmatrix},$$

$$B_2 = C_2^T = \begin{bmatrix} 0.3677 & 0.1779 \\ 0.3285 & 0.6908 \\ 0.7729 & 0.2639 \\ 0.2973 & 0.4577 \end{bmatrix}, \text{ and } D_{12} = D_{21}^T = \begin{bmatrix} 0.5369 & 0.2923 \\ 0.0665 & 0.2897 \\ 0.4939 & 0.7538 \\ 0.4175 & 0.0968 \end{bmatrix}.$$

The SPR/$\mathcal{H}_\infty$ controller is obtained using Procedure 3.25. The details of the controller parameters are available from the companion website for the book. The singular value plot of the resulting SPR/$\mathcal{H}_\infty$ controller is shown in Figure 3.22. The IFP index plot of the controller is shown in Figure 3.23. It is observed that α_i oscillates above zero at the beginning and then monotonically decreases below zero with 19 iterations. The $\mathcal{H}_\infty$-norm performance achieved by a controller without the SPR condition is $\|T_{zw}\|_\infty = 4.8$. The SPR controller achieves slightly poorer performance $\|T_{zw}\|_\infty = 6.3$ but guarantees robust stability due to the inherent passivity of the process.

3.6 Summary

The robust control approaches based on the passivity framework are presented in this chapter. They include (1) a control design method based on simultaneous IFP and OFP bounds of the uncertainties, (2) an approach that combines the passivity condition with the small gain condition in different frequency bands and (3) a passive $\mathcal{H}_\infty$ control synthesis method that achieves robust stability for any passive processes. To reduce the conservativeness of control design, frequency-dependent passivity indices are used. In the derivations of detailed algorithms, multiplicative uncertainty is assumed. However, implementation of the framework for other types of uncertainty structure can be obtained in a similar manner.

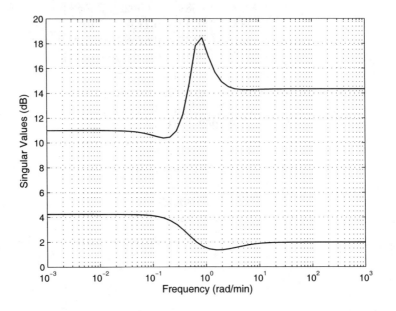

Fig. 3.22. Singular values of the SPR controller

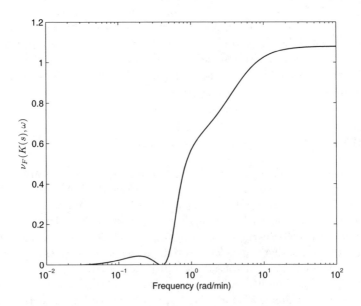

Fig. 3.23. IFP index of the SPR controller

4

Passivity-based Decentralized Control

A passive process can be stabilized by a decentralized passive controller. Therefore, the degree of passivity can imply how interactions between different loops in a multivariable process can affect the stability of the decentralized control system. In this chapter, an interaction analysis approach based on passivity is introduced. This includes steady-state and dynamic interaction analysis for both multiloop (fully decentralized) control and multi-unit (block diagonal) control. Passivity-based decentralized control design approaches are also presented.

4.1 Introduction

The linear decentralized control problem can be described as follows: Assume that the full plant is represented by an $m \times m$ rational transfer function $G(s)$ that maps the input vector $u = \begin{bmatrix} u_1, u_2, \ldots, u_m \end{bmatrix}^T$ to the output vector $y = \begin{bmatrix} y_1, y_2, \ldots, y_m \end{bmatrix}^T$:

$$
G(s) = \begin{bmatrix}
g_{11}(s) & g_{12}(s) & \cdots & g_{1m}(s) \\
\vdots & g_{22}(s) & \cdots & g_{2m}(s) \\
g_{m-1,1}(s) & \vdots & \ddots & \vdots \\
g_{m1}(s) & \cdots & g_{m,m-1} & g_{mm}(s)
\end{bmatrix}, \tag{4.1}
$$

and the diagonal (or block diagonal) system is denoted as

$$
G_d(s) = \text{diag}\left\{g_{11}(s), g_{22}(s), \cdots, g_{ii}(s), \cdots, g_{mm}(s)\right\}. \tag{4.2}
$$

The decentralized controller is a diagonal (or block diagonal) system:

$$
C(s) = \text{diag}\left\{c_1(s), \ldots, c_i(s), \ldots, c_m(s)\right\}, \tag{4.3}
$$

such that the ith controller $c_i(s)$ is designed based on model $g_{ii}(s)$ and controls y_i by manipulating u_i $(i = 1 \ldots m)$.

Decentralized control for multivariable systems is dominant in industrial process control applications due to their simplicity [112]. There are fewer communication links required and fewer controller parameters that need to be chosen, compared to the full multivariable control system. If the controller blocks are designed properly, it is generally easier to achieve better failure tolerance with a decentralized control structure. However, as each decentralized controller acts only based on the feedback in its own loop or block, the decentralized structure inevitably leads to performance deterioration due to interactions between loops and blocks. If interactions are not considered in the controller design, one may even risk instability of the closed-loop system. Therefore, interaction analysis is important in decentralized control.

The relative gain array (RGA) [23] and the block relative gain array (BRGA) [81] are commonly used interaction measures. They are simple, since only steady-state information is required, but cannot be used to infer the stabilizability of process systems using decentralized controllers (referred to as decentralized stabilizability). To overcome the above deficiencies, methods using dynamic models of the processes were developed [63, 84]. Rosenbrock [99] generalized and extended the classical Bode-Nyquist design paradigms to give conditions for diagonally decentralized stability in the gain space. Another approach to interaction analysis is based on the robust decentralized control framework [47], where the interactions were modelled as uncertainty, and an interaction measure was proposed based on structured singular values.

The passivity theory provides a different avenue toward interaction analysis. As we learned in Chapter 2, the supply rate for passive systems is defined as $w(t) = y^T(t)u(t)$ (where $y, u \in \mathbb{R}^m$). Therefore the positive real condition (as given in (2.45)),

$$\int_{t_0}^{t_1} y^T(t)u(t)dt = \int_{t_0}^{t_1} \sum_{i=1}^{m} y_i(t)u_i(t)dt \geqslant 0, \qquad (4.4)$$

actually defines the relationship between the ith output $y_i(t)$ and ith input $u_i(t)$ (rather than cross terms of $y_i(t)u_j(t)$, $i \neq j$) of the process. (This is no longer valid when the supply rate is $w(t) = y^T(t)Su(t)$ with a full matrix S.) Not surprisingly, there is a connection between passivity and decentralized control: if the process is strictly passive, it can be stabilized by *any* decentralized passive controller. Such a controller can be multiloop or block diagonal.

If the magnitude of the diagonal elements of $G(s)$ in (4.1) is significantly larger than that of the off-diagonal ones, *e.g.*,

$$|G_{ii}(j\omega)| > \sum_{j=1,\ j\neq i}^{m} |G_{ij}(j\omega)|, \quad \forall\, i, \, \forall\, \omega, \qquad (4.5)$$

then the process is said to be diagonally dominant. Obviously, diagonally dominant processes can often be effectively controlled by decentralized controllers because they have small loop interactions. While the passivity condition guarantees the decentralized stability of the decentralized control system, it does not imply diagonal dominance. For example, the following system:

$$G(s) = \begin{bmatrix} \frac{1}{s+1} & -a \\ a & \frac{2}{s+2} \end{bmatrix}, \tag{4.6}$$

with a large value of a is not diagonally dominant. However, it is easy to verify that $G(s)$ is strictly input passive for *any* a and could be stabilized by *any* passive decentralized controller. It was pointed out by Campo and Morari [26] using a counterexample that diagonal dominance is not necessary for decentralized stabilizability. It is the magnitude of interactions and also the way the subsystems interact with each other that may cause stability problems in decentralized feedback control. Passivity is related to the phase condition of the entire process system with the presence of the interaction between subsystems and loops. Therefore, it is possible to develop a passivity-based interaction analysis which indicates the destabilizing effect of the interactions.

There are two main issues in decentralized control [87]. The first is the pairing problem, *i.e.*, to decide which set of measurements (plant outputs) y_i should be controlled by which set of manipulated variables (plant inputs) u_i. Different pairings for the same plant may lead to quite different stability and performance results of decentralized control. So a pairing scheme should be chosen that is physically feasible and enables good control performance. An important concept is decentralized integral controllability (DIC). It addresses the closed-loop performance that a process can achieve under decentralized control and thus can be used in determining workable pairings. For highly coupled processes or multi-unit processes, a block decentralized control structure that consists of multiple multivariable subcontrollers should be considered. Therefore, interaction analysis for a block decentralized control structure, which is defined as block decentralized integral controllability (BDIC), is introduced [151].

The second issue in decentralized control is to design individual control loops (or blocks). This involves dynamic interaction analysis and decentralized control design that guarantees closed-loop system stability and achieves certain performance specifications.

In the following sections, sufficient conditions for DIC and BDIC are developed on the basis of passivity. The DIC condition is then extended to nonlinear systems. Then we show how the concept of passivity is used in dynamic interaction analysis and decentralized control design.

4.2 Decentralized Integral Controllability

DIC analysis determines whether a multivariable plant can be stabilized by multiloop controllers, whether the controller can have integral action to ensure zero steady-state error and whether the closed-loop system will remain stable when any subset of loops is detuned or taken out of service. DIC can be defined as follows:

Definition 4.1 (Decentralized integral controllability [114]). *A multivariable process with a transfer function $G(s) \in \mathbb{C}^{m \times m}$ is decentralized integral controllable (DIC) if there exists a decentralized controller with integral action in each loop, such that the feedback system is stable and such that each individual loop may be detuned independently by a factor ε_i $(0 \leq \varepsilon_i \leq 1,\ i = 1 \dots m)$ without introducing instability.*

DIC is a property of a given plant in combination with some prespecified control structure including manipulated/controlled variable pairings, independent of the controllers. If a system is DIC, then it is possible to achieve stable and offset-free control of the overall closed-loop system by tuning every loop separately. A different process pairing scheme will lead to a different process model $G(s)$ — when process inputs and outputs are swapped, so are the corresponding columns and rows in the transfer function $G(s)$. Clearly, an ideal pairing should lead to a process model which is DIC. A number of necessary and sufficient DIC conditions have been reported in the literature [75, 113]. The most widely used DIC analysis methods are based on interaction measures (*e.g.*, [47, 87, 113]) which imply that a system is DIC if it is generalized diagonally dominant:

Theorem 4.2 (Small gain-based sufficient condition for DIC [87]). *An LTI stable process $G(s)$ is DIC if it is generalized diagonally dominant in steady state, i.e.,*

$$\bar{\mu}\left(E_d(0)\right) < 1, \tag{4.7}$$

where $E_d(s) = [G(s) - G_d(s)]\, G_d^{-1}(s)$ and $G_d(s)$ consists of the diagonal transfer functions of $G(s)$ and $\bar{\mu}\left(E_d(0)\right)$ is the upper bound of the diagonally structured singular value of $E_d(0)$.

Necessary conditions for DIC are useful in screening out unworkable pairings and/or process designs. An easy-to-use necessary condition was derived based on the RGA:

Theorem 4.3 (Necessary condition for DIC[87]). *An $m \times m$ LTI stable process $G(s)$ is DIC only if*

$$\lambda_{ii}\left(G(0)\right) \geq 0, \quad \forall\, i = 1, \dots, m, \tag{4.8}$$

where $\lambda_{ii}\left(G(0)\right)$ is the ith diagonal element of the RGA matrix of $G(0)$.

In this section, we introduce a passivity-based DIC analysis developed by the first author and his co-workers. Based mainly on [12], we will show how the concept of passivity is linked with DIC and can be used to determine the DIC property of a given process.

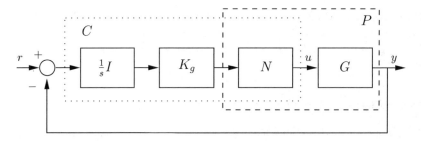

Fig. 4.1. Decentralized integral controllability

4.2.1 Passivity-based DIC Condition

As shown in Figure 4.1, the decentralized controller $C(s)$ is decomposed into $C(s) = N(s)K_g/s$, where $N(s) \in \mathbb{C}^{m \times m}$ is diagonal, stable and does not contain integral action, and $K_g = \text{diag}\{k_i\}$, $i = 1, \ldots, m$. The problem of DIC can be interpreted as whether the closed-loop system of K_g/s and system $P(s) = G(s)N(s)$ is stable and remains stable when K_g is reduced to

$$K_{g\varepsilon} = \text{diag}\{k_i \varepsilon_i\}, \ 0 \le \varepsilon_i \le 1, \ i = 1, \ldots, m. \tag{4.9}$$

Since the Passivity Theorem can deal with systems that have unlimited gain (*e.g.*, controllers with integral action), it can be used directly to analyse decentralized stability by simply examining the open-loop systems of $C(s)$ and $G(s)$. The following $m \times m$ system

$$K(s) = \frac{1}{s}K_g = \frac{1}{s} \text{diag}\{k_i\}, \ k_i \ge 0, \ i = 1, \ldots, m, \tag{4.10}$$

has m nonrepeated poles at $s = 0$. From Definition 2.24, $K(s)$ is positive real and remains positive real when the gain matrix K_g is reduced to $K_{g\varepsilon}$ as in (4.9). Therefore, the closed-loop system in Figure 4.1 will be stable if $P(s)$ is strictly positive real. As a special case ($N(s) = I$), if $G(s)$ is strictly positive real, regardless of the interactions between different channels of the process system, it can be stabilized by any positive real controllers, including $K(s)$ defined in (4.10), and this leads to the conclusion that $G(s)$ is DIC.

Many processes are not strictly positive real and thus cannot be analysed directly by using the Passivity Theorem. The conservativeness of the above passivity-based condition can be reduced by

1. rescaling the process transfer function;
2. combining the positive-real condition with the small gain condition, similar to Theorem 3.21.

If the process is stable, only the positive real condition in steady state needs to be considered for DIC, as shown in the following theorem:

Theorem 4.4 (Passivity-based sufficient condition for DIC [12]). *A stable linear multivariable process with a transfer function $G(s) \in \mathbb{C}^{m \times m}$ is DIC if a real diagonal matrix $D = \text{diag}\{d_i\}$ $(d_i \neq 0, i = 1 \ldots m)$ can be found such that*

$$G(0)D + DG^T(0) \geq 0, \tag{4.11}$$

where $G(0)$ is the steady-state gain matrix.

The proof of the above theorem was originally published in [12] and has been included in Section B.2 for completeness.

Theorem 4.4 can be seen as a special case of Theorem 3.21. It is only a sufficient condition because it is based on the positive real condition (Passivity Theorem) which itself is a sufficient stability condition for interconnected systems. The positive real condition could be conservative because it guarantees the stability of interconnected systems for *any* gain conditions. However, when one of the subsystems has infinite gain (*e.g.*, having integral action), then the positive real condition is far less conservative. The above passivity-based DIC condition is very tight and less conservative than the small gain-based condition (as shown in Example 4.6). Inequality 4.11 is actually the positive real condition on the steady-state gain matrix $G(0)$ with a rescaling matrix D. Matrix $D = \text{diag}\{d_i\}$ $(i = 1, \ldots, m)$ rescales $G(0)$ and adjusts the sign of each column of $G(0)$ such that the diagonal elements of $G(0)D$ are positive. Matrix D will then be absorbed by the decentralized controller because D is diagonal and constant. Therefore, control loop i should be reverse acting if $\text{sign}(d_i) > 0$ and direct acting if $\text{sign}(d_i) < 0$.

4.2.2 Computational Methods

Here we present two computational methods for checking the condition given in Theorem 4.4:

Semidefinite programming

A feasibility problem with a linear matrix inequality (LMI) can be set up with a real and diagonal matrix decision variable D. A matrix D which satisfies (4.11), if it exists, can be found by using the semidefinite programming (SDP) technique. The decision variable can be rewritten in the following form [21]:

$$D = U_0 + \sum_{j=1}^{q} x_j U_j, \tag{4.12}$$

where $x_j \in \mathbb{R}$, and $U_j \in \mathbb{R}^{m \times m}$ are sparse constant matrices with only one nonzero element "1" located at the position corresponding to one nonzero element of D. Therefore, the decision variable D can be written as an affine function of a number of scalar variables x_j $(j = 1, \ldots, q)$ while its prespecified diagonal structure is represented by a series of matrices U_j $(j = 1, \ldots, q)$. The LMI problem formulated above, with the structural constraint on decision variable D, is convex and can be solved by using any SDP software package, including MATLAB® *LMI Toolbox* (which is later renamed as *LMI Lab* and included in the *Robust Control Toolbox*). This approach is numerically efficient and reliable for the positive definite condition

$$G(0)D + DG^T(0) > 0, \tag{4.13}$$

but may encounter numerical problems when the minimum eigenvalue of $\left[G(0) D + DG^T(0) \right]$ is zero.

Structured singular value approach

To avoid the numerical problem in the SDP approach, we can convert the positive real condition in (4.11) into a gain condition by using the *Cayley transformation* given in (3.63) and then checking the structured singular value of the transformed steady-state gain matrix. Define a diagonal sign matrix $V = \text{sign}(G_d(0))$. The ith element of V is $+1$ (or -1) if the ith element of $G_d(0)$ is positive (or negative). Therefore $G^+(0) = G(0)V$ is the modified gain matrix such that all diagonal elements of $G^+(0)$ are positive. Because the sign matrix has been absorbed into the steady-state gain matrix, the decision variable should now be $D^+ = DV > 0$. Clearly, from Theorem 4.4, $G(s)$ is DIC if $G^+(0)$ is nonsingular and a real and positive definite diagonal matrix D can be found such that,

$$G^+(0) D^+ + D^+ \left[G^+(0) \right]^T \geq 0. \tag{4.14}$$

Following a proof similar to [26], the following proposition can be proved:

Proposition 4.5. *Define* $H = \left[I - G^+(0) \right] \left[I + G^+(0) \right]^{-1}$ *and* $F = (D^+)^{-\frac{1}{2}}$. *Inequality 4.14 holds (thus $G(s)$ is DIC) if and only if*

$$\bar{\sigma} \left\{ FHF^{-1} \right\} \leq 1, \tag{4.15}$$

where $\bar{\sigma} \left\{ \cdot \right\}$ *denotes the maximum singular value.*

The maximum singular value $\bar{\sigma} \left\{ FHF^{-1} \right\}$ is actually the upper bound of the diagonally scaled structured singular value of H. Therefore, Proposition 4.5 gives a computational method for checking the DIC condition. Procedures to calculate structured singular values can be used to test (4.15). In MATLAB® *Robust Control Toolbox*, this can be done by using the function PSV, which calculates the maximum diagonally scaled structured singular value via the Perron eigenvector approach.

Example 4.6 ([26]). Consider a system with the following transfer function:

$$G(s) = \begin{bmatrix} 1 & 0 & 2 \\ \frac{1}{s+1} & 1 & \frac{-4s}{s+1} \\ 0 & 4 & 1 \end{bmatrix}, \tag{4.16}$$

with steady-state gain matrix given by

$$G(0) = \begin{bmatrix} 1 & 0 & 2 \\ 1 & 1 & 0 \\ 0 & 4 & 1 \end{bmatrix}. \tag{4.17}$$

Here we test the DIC conditions mentioned in this section:

1. RGA-based necessary condition (Theorem 4.3): The RGA matrix of $G(0)$ is

$$\Lambda(G(0)) = \begin{bmatrix} 1/9 & 0 & 8/9 \\ 8/9 & 1/9 & 0 \\ 0 & 8/9 & 1/9 \end{bmatrix}. \tag{4.18}$$

Because all diagonal elements are positive, this process may be DIC.

2. Small gain-based sufficient condition (Theorem 4.2):

$$E_d(0) = [G(0) - G_d(0)] G_d^{-1}(0) = \begin{bmatrix} 0 & 0 & 2 \\ 1 & 0 & 0 \\ 0 & 4 & 0 \end{bmatrix}. \tag{4.19}$$

Because $\bar{\mu}(E_d(0)) = 2 > 1$, the small gain-based condition is not satisfied.

3. The strictly positive real condition in (4.13): $G(0)$ has two imaginary eigenvalues and thus there does not exist a matrix D that satisfies (4.13).

4. Passivity-based DIC condition (Theorem 4.4): Using the Cayley transformation,

$$H = \left[I - G^{+}(0)\right]\left[I + G^{+}(0)\right]^{-1}$$
$$= \begin{bmatrix} -\frac{1}{2} & 1 & -\frac{1}{2} \\ -\frac{1}{4} & -\frac{1}{2} & \frac{1}{4} \\ \frac{1}{2} & -1 & -\frac{1}{2} \end{bmatrix}. \tag{4.20}$$

It can be found that $\bar{\mu}(H) = 1$. One of the possible values of the diagonal scaling matrix is

$$F = \begin{bmatrix} 1 & 0 & 0 \\ 0 & 2 & 0 \\ 0 & 0 & 1 \end{bmatrix}. \tag{4.21}$$

Thus there exists a matrix

$$D^{+} = (FF)^{-1} = \begin{bmatrix} 1 & 0 & 0 \\ 0 & \frac{1}{4} & 0 \\ 0 & 0 & 1 \end{bmatrix},$$

such that (4.14) is satisfied. This leads to the conclusion that this process is DIC.

The above process was found to be DIC [26]. This example shows that the passivity-based DIC condition (Theorem 4.4) is less conservative than the small gain-based condition.

4.3 DIC Analysis for Nonlinear Processes

Since the concept of passivity applies to both linear and nonlinear systems, passivity-based DIC analysis can be extended to nonlinear processes. Because most chemical processes are nonlinear, the usual practice is to perform the DIC analysis based on linearised models. However, this approach could produce misleading results because they may indicate only controllability within a small neighbourhood around the operating point [89]. When dealing with processes with high nonlinearity, a DIC condition that can be used to analyse nonlinear processes is useful. In this section, we present the nonlinear version of the sufficient condition for DIC. This section is based mainly on [119].

4.3.1 DIC for Nonlinear Systems

Similar to the linear version, the DIC property of nonlinear systems indicates whether a nonlinear multivariable process can be stabilized by multiloop linear or nonlinear controllers with integral action to ensure zero steady-state error and whether the closed-loop remains stable when any subset of loops is arbitrarily detuned. As in the linear DIC analysis, we consider only "square" processes (*i.e.*, the system has the same number of inputs and outputs). Consider the following nonlinear model described by the following equations with an input vector $u \in \mathbb{R}^m$, an output vector $y \in \mathbb{R}^m$ and a state vector $x \in \mathbb{R}^n$:

$$G : \begin{cases} \dot{x} = f(x, u) \\ y = g(x, u). \end{cases} \tag{4.22}$$

The control system has the same configuration as in Figure 4.1 except that both the controller and process are nonlinear. The decentralized nonlinear controller C with integral action includes three components: the stable diagonal nonlinear controller N, the gain matrix $K_g = \mathrm{diag}\{k_i\}$ ($k_i > 0$, $i = 1, 2, \cdots, m$) and the integrator. However, different from the linear DIC analysis, the stability for nonlinear systems has to be analysed for a particular equilibrium point. It is assumed that the state $x(t)$ is uniquely determined by its initial value $x(0)$ and the input function $u(t)$.

Another assumption made for convenience is that the system (4.22) has equilibrium at the origin, that is

$$f(0, 0) = 0 \quad \text{and} \quad g(0, 0) = 0. \tag{4.23}$$

If the equilibrium x_e is not at the origin, a translation is needed by redefining the state x as $x - x_e$. The nonlinear DIC is defined as follows:

Definition 4.7 (Decentralized integral controllability for nonlinear processes [119]).

Consider the closed-loop system shown in Figure 4.1. For the nonlinear process G described by (4.22),

1. *If there exists a decentralized integral controller C, such that the unforced closed-loop system ($r = 0$) is globally asymptotically stable (GAS) for the equilibrium $x = 0$ and such that the globally asymptotic stability is maintained if each individual loop of controller C is detuned independently by a factor ε_i ($0 \leq \varepsilon_i \leq 1$, $i = 1, \cdots, m$), then the nonlinear process G is said to be decentralized integral controllable(DIC) for the equilibrium $x = 0$.*
2. *If the closed-loop system is asymptotically stable (AS) near the region of the equilibrium $x = 0$, then the nonlinear process G is said to be locally decentralized integral controllable (LDIC) around the equilibrium $x = 0$.*

From the above definition, it is obvious that the process G has to be GAS around the equilibrium $x = 0$ to be DIC. System N is generally nonlinear. If a linear diagonal system N can be found such that the closed-loop stability in the above DIC definition can be achieved, then this process is said to be DIC by using a linear controller. This is a desirable property as linear control systems usually have substantially lower design, implementation and maintenance demands than nonlinear control systems [89]. The definition of local DIC is concerned only with the vicinity of the equilibrium point, which can often be assessed by testing the linear DIC condition in (4.11) on the linearised models around a particular equilibrium point.

4.3.2 Sufficient DIC Condition for Nonlinear Processes

The sufficient DIC condition for nonlinear systems was given in [119]. Similar to the conditions for linear processes, the DIC property of a nonlinear process can be determined based on its input-output relationship at steady state.

Define $K'_{g\varepsilon} = \text{diag}\{\varepsilon_i k_i\}$ (where ε_i is the detuning factor for the ith loop) and $K_{g\varepsilon} = \eta K'_{g\varepsilon}$ (where η is a small positive real scalar). The elements of diagonal matrix $K_g = \text{diag}\{k_i\}$ are adjusted such that $k_i > 0 \ \forall \ i = 1, \ldots, m$. The sign adjustment is absorbed into the diagonal controller N. Assume that the state equation of the generalized process P, which is the serial connection of process G and the diagonal controller N (as shown in Figure 4.1), is modelled as

$$P : \begin{cases} \dot{x} = f(x, u_1) \\ y_1 = g(x, u_1). \end{cases} \tag{4.24}$$

The state equation for the linear integral controller is expressed as

$$C_l : \begin{cases} \dot{\xi} = \eta K'_{g\varepsilon} u_2 \\ y_2 = \xi. \end{cases} \tag{4.25}$$

The sufficient DIC condition for nonlinear processes can be presented as follows:

Theorem 4.8 (DIC conditions for nonlinear processes [119]). *Consider the closed-loop system in Figure 4.1. Assume that the generalized process P and the linear part of the controller C_l are described by (4.24) and (4.25), respectively. The nonlinear process G is DIC for the equilibrium $x = x_e$ if a decentralized controller N can be found such that the generalized plant P satisfies all of the following conditions:*

1. *The equation $0 = f(x, u_1)$ obtained by setting $\eta = 0$ in (4.24) implicitly defines a unique C^2 function $x = h(u_1)$ for $u_1 \in U_1 \subset \mathbb{R}^m$.*
2. *For any fixed $u_1 \in U_1 \subset \mathbb{R}^m$, the equilibrium $x_e = h(u_1)$ of the system $\dot{x} = f(x, u_1)$ is GAS and locally exponentially stable (LES) (as defined in Definition 2.2).*
3. *The steady-state input output function $g(h(u_1), u_1)$ of the generalized process P satisfies the following conditions:*

$$u_1^T g(h(u_1), u_1) > 0 \tag{4.26}$$

(when $u_1 \neq 0$ and $u_1 \in U_1$), and

$$u_1^T g(h(u_1), u_1) \geq \rho |u_1|^2 \tag{4.27}$$

(for some scalar $\rho > 0$) for u_1 in a neighbourhood of $u_1 = 0$.

The proof of the above theorem was originally published in [119] and is included in Section B.3. Although the theorem is proved using the Singular Perturbation Theorem [70] for the sake of rigorousness, the key conditions in (4.26) and (4.27) are a strictly input passivity condition on the steady-state input output function $g(h(u_1), u_1)$.

The conditions associated with Theorem 4.8 are all for the general process P (which is the serial connection of process G and the diagonal controller N) rather than the nonlinear plant G. To verify the above DIC conditions, a diagonal nonlinear controller N needs to be constructed. Searching for a linear diagonal system N is often attempted first because it is much easier to develop than a nonlinear system. In this case, the overall controller will be linear.

If we denote G_0 as the steady-state gain matrix of the *linearised model* of G around the equilibrium, and simply choose N as a real diagonal matrix D, then (4.26) and (4.27) are reduced to the existence of a real diagonal matrix $D = \text{diag}\{d_i\}$ ($d_i \neq 0$, $i = 1, \ldots, m$) such that

$$G_0 D + D G_0^T > 0, \tag{4.28}$$

which is a sufficient condition for local DIC. This is the linear DIC condition given in Theorem 4.4. If there does not exist a D matrix such that (4.28) is satisfied, then the global conditions in Theorem 4.8 do not hold. Because the local DIC condition is usually much easier to test, it can be used as a necessary condition for Theorem 4.8 (Note: Not the necessary condition for DIC because Theorem 4.8 itself is a sufficient condition for DIC).

The necessary conditions for linear processes (*e.g.*, Theorem 4.3) can be used as necessary conditions for nonlinear DIC. They are particularly useful in the nonlinear case because Theorem 4.8 could be difficult to test. A mild necessary condition of process P being DIC for both local and global around the equilibrium can be drawn from Theorem 8 in [26]. By using singular perturbation analysis [70], it can be found that the above linear necessary DIC conditions are necessary conditions for local nonlinear DIC.

4.3.3 Computational Method for Nonlinear DIC Analysis

Similar to other nonlinear analysis, a test of the DIC conditions of Theorem 4.8 analytically is often difficult and sometimes even impossible. For physical processes, the DIC analysis needs to be performed only for the operating region of interest instead of a "theoretical" global space. In this section, the computational method to assess the DIC for a given nonlinear process G : $u \rightarrow y$, within the operating region of $u \in U \subset \mathbb{R}^m$, $y \in Y \subset \mathbb{R}^m$, with respect to the steady-state equilibrium point (u_e, y_e) is as follows:

Procedure 4.9 (Numerical analysis of nonlinear DIC [119])

1. *Check the stability (GAS and LES) of the nonlinear process G (Condition 2 of Theorem 4.8). If process P is unstable, it is not DIC.*
2. *Linearise the nonlinear process G around the equilibrium point, and check whether the steady-state gain matrix $G(0)$ of the linearised model satisfies the necessary conditions for linear systems (e.g., the necessary conditions in [26, 76, 145]). If any of these necessary conditions are not satisfied, then process G is not DIC. Otherwise, proceed to the next step.*
3. *Test the sufficient condition for linear DIC in (4.11) using the linearised model around the equilibrium point, as described in Section 4.2.2. If this sufficient DIC condition is not satisfied, DIC of the nonlinear process G is not conclusive based on Theorem 4.8. Otherwise, proceed to the next step.*
4. *Check whether Condition 1 of Theorem 4.8 is satisfied by the nonlinear model (4.24) of the generalized process P : $u_1 \rightarrow y$ ($u_1 \in U_1 \subset \mathbb{R}^m$, $y \in Y \subset \mathbb{R}^m$), where P is process G in series connection with the diagonal system N. U_1 is the region of u_1 corresponding to $u \in U$ and u_{1e} is the steady-state equilibrium point corresponding to $u = u_e$. An easy and often effective choice of system N is the real diagonal constant matrix D obtained in Step 3. Solve the steady-state equation and find a unique function $x = h(u_1)$ from the equation $0 = f(x, u_1)$.*

Check whether $x = h(u_1) \in \mathbb{C}^2$ for any $u_1 \in U_1$. If not, DIC is not conclusive. Otherwise, proceed to the next step.

5. *Redefine the input Δu as $u_1 - u_{1e}$ and output Δy as $y - y_e$ such that the steady-state input output function $\Delta y = g(h(\Delta u), \Delta u)$ is unbiased in the sense that $0 = g(h(0), 0)$. Then, check Condition 3 of Theorem 4.8 for positiveness of the inner product $\Delta y^T \Delta u = \Delta u^T g(h(\Delta u), \Delta u)$ numerically in the region of interest. The condition (to ensure LES) of $\Delta u^T g(h(\Delta u), \Delta u) \geq \rho |\Delta u|^2$ (for some scalar $\rho > 0$) for Δu in a neighbourhood of $\Delta u = 0$ is guaranteed by the use of D as system N. If the inner product $\Delta y^T \Delta u$ satisfies Condition 3 in the region of interest, then process G is DIC.*

4.3.4 Nonlinear DIC Analysis for a Dual Tank System

We illustrate how to use the above DIC analysis procedure, using a dual tank level control problem. This example is taken from [119]. As shown in Figure 4.2, water flows into two tanks and is controlled by Pump 1 and Pump 2 to produce flow rates f_1 and f_2. The outlet flow rates of the two tanks are denoted f_{o1} and f_{o2}, respectively. The liquid levels in tank 1 and tank 2 are h_1 and h_2. Assume that $h_1 > h_2$ and the flow rate from tank 1 to tank 2 is f_{12}.

The liquid levels in tanks 1 and 2 can be described by the following equations:

$$\begin{aligned}
\frac{dh_1}{dt} &= \frac{1}{A}(-f_{12} - f_{o1} + f_1) = \frac{1}{A}(-k_2\sqrt{h_1 - h_2} - k_1\sqrt{h_1} + f_1), \\
\frac{dh_2}{dt} &= \frac{1}{A}(f_{12} - f_{o2} + f_2) = \frac{1}{A}(k_2\sqrt{h_1 - h_2} - k_1\sqrt{h_2} + f_2),
\end{aligned} \tag{4.29}$$

where $A = 1$ m^2, $k_1 = 0.26$ m$^{-\frac{1}{2}}$/min and $k_2 = 0.13$ m$^{-\frac{1}{2}}$/min. If we select the flow rates f_1 and f_2 as control inputs and the heights of the liquid level

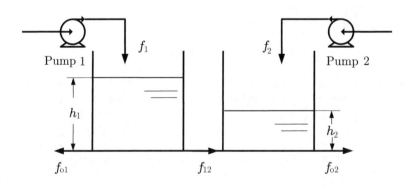

Fig. 4.2. Schematic diagram of a dual tank level control process

h_1 and h_2 as control outputs, then the process model in (4.29) can be written as

$$\dot{x}_1 = \frac{1}{A}(-k_2\sqrt{x_1 - x_2} - k_1\sqrt{x_1} + u_1),$$

$$\dot{x}_2 = \frac{1}{A}(k_2\sqrt{x_1 - x_2} - k_1\sqrt{x_2} + u_2), \qquad (4.30)$$

$$y_1 = x_1,$$

$$y_2 = x_2.$$

The process model in (4.30) is found to be LES from its linearised model around the equilibrium. This model is also GAS in the region of interest (details can be found in [119]):

$$h_1 > 0, \ h_2 > 0 \text{ and } h_1 - h_2 > 0. \qquad (4.31)$$

The DIC property of this process around an equilibrium point

$$x_e = [h_{1e}, \ h_{2e}]^T = [8.8, \ 5.8]^T \text{m}, \qquad (4.32)$$

is analysed. It can be verified that (4.11) is satisfied with $D = I$. Therefore, the linearised system is DIC. The conditions in Theorem 4.8 for the nonlinear model in (4.30) are analysed with $D = I$.

By redefining the state, input and output variables as

$$\tilde{x} = x - x_e = [x_1 - 8.8, \ x_2 - 5.8]^T \text{ m},$$

$$\Delta u = u_1 - u_e = [u_{11} - 1, \ u_{12} - 0.4]^T \text{ m}^3/\text{min}, \qquad (4.33)$$

$$\Delta y = y - y_e = [y_1 - 8.8, \ y_2 - 5.8]^T \text{ m}^3/\text{min},$$

the steady-state input output mapping can be found as follows:

$$\Delta y_1 = (25\sqrt{5(1 + \Delta u_1)^2 + 2(1 + \Delta u_1)(2/5 + \Delta u_2) - 3(2/5 + \Delta u_2)^2}$$
$$- 135 - 125\Delta u_1 - 25\Delta u_2)^2/676 - 8.8,$$
$$\Delta y_2 = (25\sqrt{5(1 + \Delta u_1)^2 + 2(1 + \Delta u_1)(2/5 + \Delta u_2) - 3(2/5 + \Delta u_2)^2} \qquad (4.34)$$
$$+ 5 - 25\Delta u_1 + 75\Delta u_2)^2/676 - 5.8.$$

Condition 3 in Theorem 4.8 is reduced to $\Delta y^T \Delta u > 0$, which can be verified numerically using discrete points in the region of interest. A three-dimensional plot is given in Figure 4.3, from which it can be seen that the DIC conditions in Theorem 4.8 are satisfied in the following region:

$$\Delta u_1 \in [-0.4, \ 0.4] \text{ m}^3/\text{min} \ (f_1 \in [0.6, \ 1.4] \text{ m}^3/\text{min}), \qquad (4.35)$$
$$\Delta u_2 \in [-0.2, \ 0.2] \text{ m}^3/\text{min} \ (f_2 \in [0.2, \ 0.6] \text{ m}^3/\text{min}).$$

Therefore, this nonlinear process is DIC in the above input space.

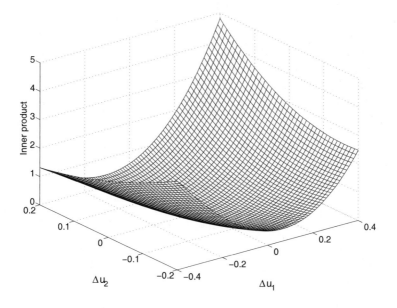

Fig. 4.3. Values of $\Delta y^T \Delta u$ [119]

4.4 Block Decentralized Integral Controllability

When process interactions are severe, the plantwide process may not be effectively controlled by fully decentralized (multiloop) controllers. In this case, a multivariable controller is often implemented for each subsystem, which can be either a part of a process unit or a subsystem of the plantwide process containing several process units. The entire control system will have a block diagonal structure. Design and implementation of such a control system is often simpler and more fault-tolerant than the full multivariable controller for the entire plantwide process [77]. Similar to the multiloop design, it is preferable that the block diagonal controller has integral action to achieve offset-free control (block integral controllability), and any one or more controller blocks can be arbitrarily detuned without endangering closed-loop stability (block detunability). We can extend the concept of DIC and define processes that can be controlled by the above block diagonal control systems to be block decentralized integral controllable (BDIC). This section is based mainly on [151].

Consider a feedback system under block decentralized control, as shown in Figure 4.4. The $m \times m$ process transfer function $G(s)$ is partitioned in the following block form:

$$G(s) = [G_{ij}(s)]_{i,j=1,\ldots,k}, \tag{4.36}$$

where each block $G_{ij}(s)$ is an $m_i \times m_j$ transfer function submatrix. Define

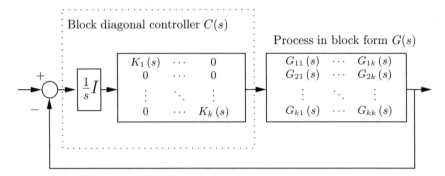

Fig. 4.4. Feedback system under block diagonal control

$$G_b(s) = \text{diag}\,[G_{ii}(s)]_{i=1,\ldots,k}\,.\qquad(4.37)$$

Under block decentralized control, the diagonal blocks $G_{ii}(s)$ are controlled by *multivariable* subcontrollers that form the overall controller $C(s)$ with the following block diagonal structure:

$$C(s) = \text{diag}\,[C_i(s)]_{i=1,\ldots,k}\,,\qquad(4.38)$$

where $C_i(s)$ is an $m_i \times m_i$ multivariable controller for the ith block $G_{ii}(s)$. The property of BDIC can be defined as follows:

Definition 4.10 (Block decentralized integral controllability [151]).
For a given multivariable stable process system with a transfer function $G(s) \in \mathbb{C}^{m\times m}$ in the block partition form of (4.36), if there exists a corresponding block diagonal controller in the form of (4.38) with integral action, i.e.,

$$C(s) = \text{diag}\,[C_i(s)]_{i=1,\ldots,k} = \frac{1}{s}\,\text{diag}\,[K_i(s)]_{i=1,\ldots k}\,,\qquad(4.39)$$

such that the feedback system is stable and such that each individual controller block $C_i(s)$ ($i = 1,\ldots,k$) may be detuned independently by a factor ξ_i, ($0 \le \xi_i \le 1, \forall\, i = 1,\ldots,k$), without introducing instability, this process is said to be block decentralized integral controllable (BDIC) with respect to the prespecified controller structure.

The block decentralized integral controllability represents some preferable features of a process with respect to the prespecified controller structure: (1) integral action can be used in each subcontrol system to achieve offset-free control; (2) each controller block can be detuned or switched off independently without endangering the closed-loop stability.

A concept relevant to BDIC is decentralized closed-loop integrity (DCLI), which is defined as follows:

Definition 4.11 (Decentralized closed-loop integrity [31]). *A multivariable stable process $G(s)$ is said to possess decentralized closed-loop integrity (DCLI) if it can be stabilized by a stable block diagonal controller $C(s)$ which contains integral action and if it remains stable after failure occurs in one or more control blocks.*

DCLI addresses the issue of closed-loop system stability under control subsystem failures, *i.e.*, the process can be stabilized by a controller $\hat{C}(s) = \Xi C(s)$, where

$$\Xi = \mathrm{diag}\left\{\xi_1, \ldots, \xi_i, \ldots, \xi_k\right\}, \ \xi_i \in \{0, 1\}, \ \forall \ i = 1, \ldots, k, \tag{4.40}$$

which is a special case of BDIC. For BDIC processes, the closed-loop stability remains when controller blocks are arbitrarily detuned. Therefore, a BDIC process must also be DCLI.

4.4.1 Conditions for BDIC

Sufficient condition based on the Passivity Theorem

Since the Passivity Theorem is applicable to both decentralized and multivariable systems, the passivity-based DIC condition given in Theorem 4.4 can be extended to BDIC as follows:

Theorem 4.12 (Sufficient condition for BDIC[151]). *A multivariable stable process with a transfer function $G(s) \in \mathbb{C}^{m \times m}$ is block decentralized integral controllable with respect to the prespecified controller structure in (4.38) if a nonsingular real block diagonal matrix,*

$$W = \mathrm{diag}\{W_1, \ldots, W_i, \ldots, W_k\} \in \mathbb{R}^{m \times m}, \tag{4.41}$$

where $W_i \in \mathbb{R}^{m_i \times m_i}$, can be found such that

$$G(0)W + W^T G^T(0) \geq 0. \tag{4.42}$$

Proof. Any nonsingular square block diagonal matrix W given above can be factorized into two nonsingular, square matrices M and N with the same block diagonal structure of W, *i.e.*,

$$W = MN^{-1}, \tag{4.43}$$

where $M = \mathrm{diag}\{M_i\}$, $N = \mathrm{diag}\{N_i\}$, $i = 1, \ldots, k$, M, $N \in \mathbb{R}^{m \times m}$ and M_i, $N_i \in \mathbb{R}^{m_i \times m_i}$. Thus (4.42) can be written as

$$G(0)MN^{-1} + N^{-T}M^T G^T(0) \geq 0, \tag{4.44}$$

which is equivalent to

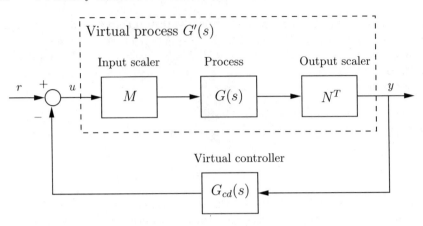

Fig. 4.5. BDIC analysis

$$N^T \left[G(0)MN^{-1} + N^{-T}M^TG^T(0) \right] N \geq 0, \tag{4.45}$$

i.e.,

$$N^TG(0)M + M^TG^T(0)N \geq 0. \tag{4.46}$$

As shown in Figure 4.5, define a virtual process $G'(s) = N^TG(s)M$. It is easy to see that $G'(0)$ is nonsingular. Inequality 4.46 indicates that $G'(0)$ satisfies the following inequality:

$$G'(0)I_m + I_mG'^T(0) \geq 0, \tag{4.47}$$

where I_m is an $m \times m$ identity matrix. From Theorem 4.4, the virtual process $G'(s)$ is DIC. As a result, there exists a multiloop controller,

$$C_d(s) = \frac{1}{s} \, \text{diag} \left\{ K_{d1}(s), \ldots, K_{dj}(s), \ldots, K_{dm}(s) \right\}, \tag{4.48}$$

which stabilizes $G'(s)$ and maintains closed-loop stability when each diagonal subcontroller is independently detuned by an arbitrary factor of $\xi_{dj} \in [0,1]$, $\forall \, j = 1, \ldots, m$. By choosing the detuning factors (ξ_i for the ith controller block, $i = 1, \ldots, k$), it can be seen that process $G'(s)$ is also stabilized by

$$\hat{C}_d(s) = \varXi C_d(s), \tag{4.49}$$

where

$$\varXi = \text{diag} \left\{ \xi_1 I_1, \ldots, \xi_i I_i, \ldots, \xi_k I_k \right\}, \; I_i \in R^{m_i \times m_i}, \xi_i \in [0,1], \forall \, i = 1, \ldots, k. \tag{4.50}$$

Therefore, the real controller "seen" by the original process $G(s)$ is

$$C(s) = MC_d(s)N^T = \frac{1}{s} \, \text{diag} \{ K_1(s), \ldots, K_i(s), \ldots, K_k(s) \}, \tag{4.51}$$

which can be detuned to

$$\hat{C}(s) = \Xi C(s) = \frac{1}{s} \operatorname{diag}\{\xi_1 K_1(s), \dots, \xi_i K_i(s), \dots, \xi_k K_k(s)\}, \qquad (4.52)$$

without causing closed-loop instability. Therefore, process $G(s)$ is BDIC.

For a given process $G(s)$, (4.42) becomes an LMI in the decision variable matrix W with a certain block diagonal structure that can be written in a form similar to (4.12). This LMI problem can be solved by using the MATLAB® *Robust Control Toolbox.*

Necessary conditions

Obviously, the following necessary conditions for DCLI are also necessary conditions for BDIC:

Theorem 4.13 (Necessary conditions for DCLI [31]). *Given a stable multivariable process with a transfer function $G(s)$, which can be partitioned into $k \times k$ block subsystems, $G(s)$ possesses decentralized closed-loop integrity only if*

1. *The following block relative gain (BRG) condition is satisfied:*

$$\det\left[\Lambda_i\left(G\left(0\right)\right)\right] > 0, \forall\ i = 1, \dots, k, \qquad (4.53)$$

   where $\Lambda_i\left(G(0)\right)$ is the BRG of $G_{ii}(s)$ [81].
2. *The Niederlinski Index (NI) [88] is positive, i.e.,*

$$NI\left[G(0)\right] = \frac{\det\left[G\left(0\right)\right]}{\det\left[G_b\left(0\right)\right]} > 0. \qquad (4.54)$$

4.4.2 Pairing Based on BDIC

BDIC is a less restrictive condition than DIC, but a more restrictive condition than DCLI. The necessary conditions in Theorem 4.13 can be used to screen out unworkable pairings before examining the sufficient conditions given in Theorem 4.12.

For a given process, it is possible to have multiple pairing schemes all of which satisfy the BDIC conditions. Under this circumstance, the best pairing scheme should be determined in conjunction with other pairing criteria. One of the major considerations is resilience, which is the quality of regulatory and servo behaviors that can be obtained by feedback control [86]. This can be quantified by using the minimum singular values (at steady state or over a frequency band) of each subsystem of $G_b\left(s\right)$. The larger the minimum singular values, the more resilient the subsystems are, because larger disturbances can be handled by the controller for given constraints on the manipulated variables.

4.4.3 BDIC Analysis of the SFE Process

In this section, we apply the BDIC conditions presented in Section 4.4.1 to analyse the control schemes for a supercritical fluid extraction (SFE) process [152]. As depicted in Figure 4.6, the SFE process consists of three physical units: extractor, stripper + reboiler and trim-cooler. Due to the high coupling between the controlled variables and varying response times in different units, the SFE process has been difficult to control.

In this process, the manipulated variables are

$$u_1 = \text{solvent flow rate (extractor)},$$
$$u_2 = \text{reflux flow rate (stripper)},$$
$$u_3 = \text{boilup rate (stripper), and}$$
$$u_4 = \text{shell-tube temperature (trim-cooler)}.$$

The controlled variables are

$$y_1 = \text{raffinate composition (extractor)},$$
$$y_2 = \text{overhead composition (stripper)},$$
$$y_3 = \text{bottoms composition (stripper), and}$$
$$y_4 = \text{solvent temperature (trim-cooler)}.$$

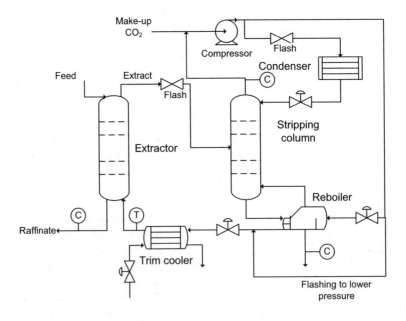

Fig. 4.6. Supercritical fluid extraction process

Here the following linearised model of the SFE process derived by Samyudia *et al.* around a steady-state operating condition is used for the BDIC analysis (details of the model and the nominal steady-state operating condition can be found in [104]):

$$\dot{x} = Ax + Bu,$$
$$y = Cx,$$

(4.55)

where $u = [u_1, u_2, u_3, u_4]^T$, $y = [y_1, y_2, y_3, y_4]^T$ and

$$A = \begin{bmatrix}
-27.1 & 8.27 & -0.945 & -0.32 & -55.11 & 0 & 0 & 0 & 0 & 0.0003 \\
15.1 & -23.4 & 9.17 & -0.945 & 0 & 0 & 0 & 0 & 0 & 0.0001 \\
0 & 15.1 & -23.4 & 8.23 & 0 & 0 & 0 & 0 & 0 & 0.0003 \\
0 & 0 & 15.1 & -27.1 & 0 & 0 & 0 & 0 & 0 & 0.0003 \\
0 & 0 & 0 & 0 & -136.1 & 136.5 & 0 & 0 & 0 & 0 \\
0 & 0 & 0 & -60.5 & -29.11 & -141.4 & 0.264 & 0 & 0 & 0.0030 \\
0 & 0 & 0 & 0 & -0.0105 & 4.96 & -5.217 & 0.264 & 0 & 0 \\
0 & 0 & 0 & 0 & -0.0009 & -0.0035 & 1.65 & -1.66 & 0 & 0 \\
0 & 0 & 0 & 0 & 0 & 0 & 0 & 0 & -41.4 & 0 \\
0 & 0 & 0 & 0 & 6.33 & 0 & 0 & 0 & 0.141 & -13.4
\end{bmatrix},$$

$$B = \begin{bmatrix}
-0.0016 & 6.4569 & -6.4568 & 0 \\
-0.0036 & 0 & 0 & 0 \\
-0.0066 & 0 & 0 & 0 \\
-0.0083 & 0 & 0 & 0 \\
-0.0577 & 1.1663 & 0.5669 & 0 \\
-0.0023 & 0.0656 & -0.0181 & 0 \\
0 & 0.0059 & -0.0958 & 0 \\
0 & 0.0359 & -0.5981 & 0 \\
0 & 0 & -15.6185 & 0 \\
0.0005 & -0.7417 & 0.7416 & 13.1095
\end{bmatrix},$$

$$C = \begin{bmatrix}
1 & 0 & 0 & 0 & 0 & 0 & 0 & 0 & 0 & 0 \\
0 & 0 & 0 & 0 & 1 & 0 & 0 & 0 & 0 & 0 \\
0 & 0 & 0 & 0 & 0 & 0 & 0 & 1 & 0 & 0 \\
0 & 0 & 0 & 0 & 0 & 0 & 0 & 0 & 0 & 1
\end{bmatrix}.$$

Using the proposed BDIC conditions, different control strategies for the SFE process are analysed. As shown in Table 4.1, the control modes of DAU1, DAU2, DAU3, DAU5 and DAU6 are BDIC and thus they are promising control schemes. In contrast, Modes SA, ESA and DAU4 are not BDIC and therefore not preferred. This conclusion is consistent with that drawn by other researchers [104] based on fairly complicated performance analysis involving dynamic simulation.

It is interesting to see that modes SA and ESA, which represent the control structures corresponding to plant physical decomposition, are not suitable. This is due to the strong coupling between the extractor and stripper units, as confirmed by Samyudia *et al.* [104]. From this example, it can be seen that

Table 4.1. BDIC-based analysis for the SFE process

Mode	Control Structure	(4.42) holds?	(4.53) holds?	(4.54) holds?	BDIC ?
SA	1. Extractor 2. Stripper+reboiler 3. Trim-cooler	No	No	No	No
ESA	1. Extractor+trim-cooler 2. Stripper+reboiler	No	No	No	No
DAU1	1. Extractor+stripper+reboiler 2. Trim-cooler	Yes	Yes	Yes	Yes
DAU2	1. Extractor+reboiler 2. Stripper+trim-cooler	Yes	Yes	Yes	Yes
DAU3	1. Extractor+stripper+trim-cooler 2. Reboiler	Yes	Yes	Yes	Yes
DAU4	1. Extractor 2. Stripper+reboiler+trim-cooler	No	No	No	No
DAU5	1. Extractor+stripper 2. Reboiler+trim-cooler	Yes	Yes	Yes	Yes
DAU6	1. Extractor+stripper 2. Reboiler 3. Trim-cooler	Yes	Yes	Yes	Yes

intuitive block diagonal control schemes based on plant physical decomposition should not always be used because they may lead to control systems with poor achievable performance.

4.5 Dynamic Interaction Measure

4.5.1 Representing Dynamic Interactions

In the previous sections, we studied several steady-state interaction measures. They are relatively simple to use but do not indicate the impact of interactions on dynamic control performance under decentralized control. One of the ways to investigate dynamic performance is to represent the interactions as process uncertainty and study their impact on closed-loop system performance and stability using the robust control framework [87].

For a system described by a full model $G(s)$ with a diagonal submodel $G_d(s) = \text{diag}\{G_{ii}\}$, the ith output y_i can be expressed as follows:

$$y_i(s) = g_{ii}(s)u_i(s) + \sum_{j=1, j \neq i}^{m} g_{ij}(s)u_j(s). \tag{4.56}$$

The first item of the above equation is the diagonal subsystem and the second item denotes the interaction between the ith channel and all other channels.

Because the ith decentralized controller $c_i(s)$ is designed only based on the model $g_{ii}(s)$, the second term can be considered the perturbation. Thus, the diagonal system model is used as the nominal model and the interaction is represented as uncertainty. Therefore, the analysis of interactions and the closed-loop stability under decentralized control can be conducted systematically using robust control theory. This approach was first developed by Grosdider and Morari [47]. The interaction can be modelled as either additive or multiplicative uncertainty. Figure 4.7 illustrates how to characterize the interaction as additive uncertainty:

$$\Delta_A = G(s) - G_d(s). \tag{4.57}$$

Then the system "seen" by the off-diagonal system is

$$M_A(s) = C(s)\left[I + G_d(s)C(s)\right]^{-1}. \tag{4.58}$$

Note that both the controller $C(s)$ and $G_d(s)$ are diagonal; therefore, system $M(s)$ is also diagonal:

$$M_A(s) = \text{diag}\left\{m_{A,1}(s), \cdots, m_{A,i}(s), \cdots, m_{A,m}(s)\right\}, \tag{4.59}$$

where

$$m_{A,i}(s) = \frac{c_i(s)}{1 + g_{ii}(s)c_i(s)}, \quad i = 1, \ldots, m.$$

From the Small Gain Theorem, the closed-loop system is stable if $M_A(s)$ and $\Delta_A(s)$ are stable and

$$\bar{\sigma}(M_A(j\omega)) < \bar{\sigma}^{-1}(\Delta_A(j\omega)), \quad \forall\, \omega \in \mathbb{R}. \tag{4.60}$$

This analysis of the interaction could be conservative due to its treatment of actually known off-diagonal systems as uncertainty. However, this approach is useful due to the following advantages:

1. The interaction measure is much simpler and the stability condition is easier to apply than other approaches.
2. The "true" model uncertainties can be dealt with within the same framework. Thus, decentralized controllers can be made robust.

The conservativeness of this approach can be reduced by taking the diagonal structure of the closed-loop system $M_A(s)$ into account. The diagonally structured singular value (μ) can be used as the interaction measure instead of the maximum singular values. Inequality 4.60 can be replaced by the following inequality [47]:

$$\bar{\sigma}(M_A(j\omega))\,\mu(\Delta_A(j\omega)) < 1, \quad \forall\, \omega \in \mathbb{R}, \tag{4.61}$$

which is less conservative because it exploits the diagonal structure of $M_A(s)$.
Interactions can also be modelled as multiplicative uncertainty:

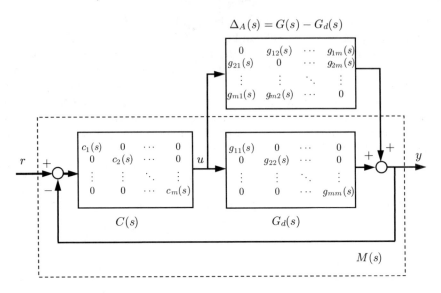

Fig. 4.7. Representing interaction as additive uncertainty

$$\Delta_M = [G(s) - G_d(s)] G_d^{-1}(s), \tag{4.62}$$

with

$$M_M(s) = G_d(s) C(s) [I + G_d(s) C(s)]^{-1}. \tag{4.63}$$

In this case, the closed-loop system is stable if $M_M(s)$ and $\Delta_M(s)$ are stable and

$$\bar{\sigma}(M_M(j\omega)) \mu(\Delta_M(j\omega)) < 1, \quad \forall\, \omega \in \mathbb{R}. \tag{4.64}$$

The values $\mu(\Delta_A(j\omega))$ and $\mu(\Delta_M(j\omega))$ are called structured singular value-based interaction measures (SSV-IM). Unfortunately, none of the above stability conditions use the phase information of the off-diagonal system and thus they can still be very conservative. In the next subsection, we adopt a passivity-based interaction measure. The basic idea is to characterize the off-diagonal system in terms of the passivity index developed in Chapter 3 to exploit both phase and gain information and derive the stability conditions for decentralized control based on the passivity and sector stability theorem.

4.5.2 Passivity-based Interaction Measure

In the development of the passivity-based interaction measure, we adopt the robust control framework similar to [47] but use the frequency-dependent IFP index (ν_-) which was introduced in Definition 3.1 to characterize the interaction. The interactions are modelled as uncertainty Δ, in the form of either

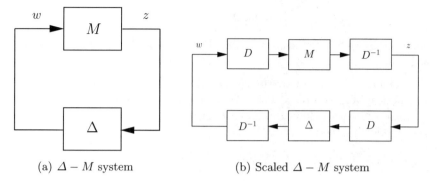

(a) $\Delta - M$ system   (b) Scaled $\Delta - M$ system

Fig. 4.8. Passivity-based interaction measure

$\Delta_A(s)$ or $\Delta_M(s)$ given in (4.57) and (4.62), respectively. Denote the closed-loop system of $G_d(s)$ and the decentralized controller $C(s)$ as system M ($M_A(s)$ or $M_M(s)$ for additive uncertainty and multiplicative uncertainty, respectively). The closed-loop system of the actual process $G(s)$ and the controller can be represented in Figure 4.8a.

Interaction measures

The interactions can be quantified by their IFP index $\nu_-(\Delta(s), \omega)$. To provide a "tighter" interaction measure, the input-output diagonal scaling similar to that used to calculate structured singular values (as in [101]) can be implemented to take into account the diagonal structure. Define D as a diagonal and nonsingular matrix. The original $\Delta - M$ system shown in Figure 4.8a is stable if and only if the feedback system with the scaling matrix D shown in Figure 4.8b is stable. Because system $M(s)$ is diagonal, $M = D^{-1}M(s)D$. However, the passivity index of $D^{-1}\Delta(s)D$ can be significantly reduced by choosing an appropriate D matrix. Because the passivity index ν_- is defined as frequency-dependent, frequency-dependent diagonal scaling matrices $D(\omega)$ can be found to minimize the passivity indices at different frequencies (similar to [16]). To facilitate the stability analysis, we define

$$\tilde{\Delta}(s) = \Delta(s)V, \tag{4.65}$$

where V is the diagonal sign matrix whose elements are determined such that the diagonal elements of $\tilde{\Delta}(s)$ are positive at steady state. For a given stable system $\Delta(s) \in \mathbb{C}^{m \times m}$, the problem of diagonal scaling of the passivity index at frequency ω can be described as follows:

Problem 4.14.

$$\min_{D(\omega)} \{t\} \tag{4.66}$$

subject to

$$D(\omega)^{-1}\tilde{\Delta}(j\omega)D(\omega) + D(\omega)\tilde{\Delta}^*(j\omega)D(\omega)^{-1} + tI > 0, \qquad (4.67)$$

where $D(\omega) \in \mathbb{R}^{m \times m}$ is a nonsingular diagonal matrix and t is a real scalar variable.

Problem 4.14 cannot be solved directly by an SDP solver because (4.67) is nonlinear and complex. This problem can be converted into a real LMI problem as shown below [16]. Since $D(\omega)$ is nonsingular, (4.67) is equivalent to the following inequality:

$$D(\omega)\left[D(\omega)^{-1}\tilde{\Delta}(j\omega)D(\omega) + D(\omega)\tilde{\Delta}^*(j\omega)D(\omega)^{-1}\right]D(\omega)$$
$$+ tD(\omega)D(\omega) > 0. \quad (4.68)$$

Define

$$H(\omega) = D(\omega)D(\omega) > 0. \qquad (4.69)$$

Then,

$$\tilde{\Delta}(j\omega)H + H\tilde{\Delta}^*(j\omega) + tH > 0. \qquad (4.70)$$

Assume that $\tilde{\Delta}(j\omega) = X(\omega) + jY(\omega)$, where both $X(\omega)$ and $Y(\omega)$ are real matrices. This leads to

$$-\left[X(\omega)H + HX^T(\omega)\right] - j\left[Y(\omega)H - HY^T(\omega)\right] - tH < 0. \qquad (4.71)$$

The above inequality holds if and only if

$$\begin{bmatrix} -X(\omega)H - HX^T(\omega) & Y(\omega)H - HY^T(\omega) \\ -Y(\omega)H + HY^T(\omega) & -X(\omega)H - H^T X(\omega) \end{bmatrix} - t\begin{bmatrix} H & 0 \\ 0 & H \end{bmatrix} < 0. \qquad (4.72)$$

Therefore, Problem 4.14 can be converted into the following generalized eigenvalue problem with constraints described in real matrix inequalities.

Problem 4.15.

$$\min_{H(\omega)} \{t\} \qquad (4.73)$$

subject to (4.72) and (4.69).

For each frequency ω, a real matrix of $H(\omega)$ can be obtained by solving the above optimization problem using an SDP solver. The passivity-based interaction measure can then be defined as

Definition 4.16 (Passivity-based interaction measure (PB-IM)). *Consider an LTI system with a transfer function $G(s)$. The diagonal elements of $G(s)$ form a diagonal subsystem $G_d(s)$. Denote $\Delta(s)$ as the representation of the interaction, in either of the following two forms: $\Delta_A(s) = G(s) - G_d(s)$ and $\Delta_M(s) = [G(s) - G_d(s)]G_d^{-1}(s)$. If $\Delta(s)$ is stable, the passivity-based interaction measure (PB-IM) at frequency ω is defined as*

$$\nu_I(G(s), \omega) \triangleq \min_D \left\{ \max\left[-\frac{1}{2}\lambda\left(D^{-1}\tilde{\Delta}(j\omega)D + D\tilde{\Delta}^*(j\omega)D^{-1}\right), 0\right] \right\}, \qquad (4.74)$$

where $\tilde{\Delta}(j\omega)$ is defined by (4.57), (4.62) and (4.65).

Matrix D is a frequency-dependent decision variable that can be obtained by solving Problem 4.15. In the rest of the book, we denote $\nu_{IA}\left(G\left(s\right),\omega\right)$ and $\nu_{IM}\left(G\left(s\right),\omega\right)$ as the interaction measures based on additive and multiplicative uncertainty models (*i.e.*, $\Delta_A(s)$ and $\Delta_M(s)$), respectively.

The above passivity-based interaction measure has the same properties as the frequency-dependent IFP index given in Property 3.3. The following decentralized stability condition can be obtained immediately from Proposition 3.4:

Theorem 4.17. *Consider an LTI process system with a transfer function $G(s)$ and a stable and minimum phase transfer function $W(s)$.*

1. *Assume that $\nu_{IA}(G(s),\omega) \leq \nu_F(W(s),\omega)$. The process can be stabilized by a decentralized controller $C\left(s\right)$ if $M_A(s)' = M_A(s)[I - W(s)M_A(s)]^{-1}$ is strictly positive real (i.e., stable and strictly input feedforward passive), where*

$$M_A\left(s\right) = C\left(s\right)\left[I + G_d\left(s\right)C\left(s\right)\right]^{-1}. \tag{4.75}$$

2. *Assume that $\nu_{IM}(G(s),\omega) \leq \nu_F(W(s),\omega)$. The process can be stabilized by a decentralized controller $C\left(s\right)$ if $M_M(s)' = M_M(s)[I - W(s)M_M(s)]^{-1}$ is strictly positive real, where*

$$M_M\left(s\right) = G_d\left(s\right)C\left(s\right)\left[I + G_d\left(s\right)C\left(s\right)\right]^{-1}. \tag{4.76}$$

Similarly, the sector-bounded passivity index (simultaneous IFP and OFP) given in Definition 3.8 can also be used as an interaction measure:

Definition 4.18 (Sector-based interaction measure (SB-IM)). *Consider an LTI system with a transfer function $G\left(s\right)$. The diagonal elements of $G\left(s\right)$ form a diagonal subsystem $G_d\left(s\right)$. If $\Delta_A\left(s\right)$ and $\Delta_M\left(s\right)$ are stable, the sector-based interaction measure (SB-IM) at frequency ω is defined as*

$$\nu_{sIA}\left(G\left(s\right),\omega,b\right) \triangleq \max\left(\nu_{S-}\left(\Delta_A(s),\omega,b\right),0\right), \tag{4.77}$$

$$\nu_{sIM}\left(G\left(s\right),\omega,b\right) \triangleq \max\left(\nu_{S-}\left(\Delta_M(s),\omega,b\right),0\right). \tag{4.78}$$

Similar to the PB-IM, the interaction can be scaled during the computation of the SB-IM. However, direct scaling could be fairly difficult. Because the values of PB-IM and SB-IM are close to each other when b is large, it is often sufficient to use the same scaling matrices optimized for the PB-IM. The SB-IM can be calculated from the PB-IM using Theorem 3.9. The definition of SB-IM immediately leads to the decentralized stability condition based on Theorem 3.11:

Theorem 4.19. *Consider an LTI process system with a transfer function $G(s)$ and a stable and minimum phase transfer function $W(s)$.*

1. *Assume that* $\nu_{sIA}(G(s), \omega, b) \leq -\nu_{S-}(W(s), \omega, b_w)$. *The process can be stabilized by a decentralized controller* $C(s)$ *if*

$$M'_A(s) = M_A(s)\left[I - W(s)M_A(s)\right]^{-1} + \frac{1}{b + b_w}I \qquad (4.79)$$

is strictly positive real (or extended strictly positive real), where $M_A(s)$ *is given in (4.75).*

2. *Assume that* $\nu_{sIM}(G(s), \omega, b) \leq -\nu_{S-}(W(s), \omega, b_w)$. *The process can be stabilized by a decentralized controller* $C(s)$ *if*

$$M'_M(s) = M_M(s)\left[I - W(s)M_M(s)\right]^{-1} + \frac{1}{b + b_w}I \qquad (4.80)$$

is strictly positive real (or extended strictly positive real), where $M_M(s)$ *is given in (4.76).*

Conceptually, the PB-IM is a "cleaner" interaction measure based on IFP. The SB-IM is theoretically less conservative and can be useful if the robust control design approach given in Section 3.3.4 is employed in decentralized control design.

Achievable decentralized control performance

Like the passivity-based uncertainty measure, the passivity-based interaction measure also implies the performance achievable by a decentralized control. For example, if $\nu_{IM}(G(s), \omega)$ is known, Theorem 4.17 implies that

$$\bar{\sigma}\left(M_M(j\omega)\right) < \frac{1}{\nu_{IM}(G(s), \omega)}, \quad \forall \, \omega. \qquad (4.81)$$

Denote

$$M_M(s) = \text{diag}\left\{m_{M,1}(s), \cdots, m_{M,i}(s), \cdots, m_{M,m}(s)\right\}, \qquad (4.82)$$

where

$$m_{M,i}(s) = \frac{g_{ii}(s)c_i(s)}{1 + g_{ii}(s)c_i(s)}, \quad \forall \, i = 1, \dots, m. \qquad (4.83)$$

Therefore,

$$|m_{M,i}(j\omega)| < \frac{1}{\nu_{IM}(G(s), \omega)}, \quad \forall \, \omega, \; i = 1, \dots, m. \qquad (4.84)$$

As a result, it is possible to make the sensitivity function of each loop $|S_i(j\omega)| = |1 - m_{M,i}(j\omega)|$ $(i = 1, \dots, m)$ arbitrarily small at frequencies when $\nu_{IM}(G(s), \omega) < 1$ but

$$|S_i(j\omega)| \geq \left|1 - \frac{1}{\nu_{IM}(G(s), \omega)}\right|, \qquad (4.85)$$

when $\nu_{IM}(G(s),\omega) > 1$. The above result is comparable to the small gain-based decentralized control condition given in (4.61) and (4.64). For example, if multiplicative uncertainty is used,

$$\bar{\sigma}\left(M_M\left(j\omega\right)\right) < \mu^{-1}\left(\Delta_M\left(j\omega\right)\right), \quad \forall\, \omega, \tag{4.86}$$

which implies that the smallest sensitivity function of each loop is bounded by

$$\left|S_i\left(j\omega\right)\right| \geq \left|1 - \mu^{-1}\left(\Delta_M\left(j\omega\right)\right)\right|, \quad \forall\, \omega, \; i = 1,\ldots,m, \tag{4.87}$$

when $\mu\left(\Delta_M\left(j\omega\right)\right) > 1$.

One special performance consideration is offset-free control. To achieve zero steady-state error of the closed-loop system using a decentralized controller, it is essential to implement an integrator in each of the single-loop controllers. Clearly, if the process $G\left(s\right)$ is stable, the sufficient condition for integral controllability is

$$\nu_{IM}(G(s),0) < 1, \tag{4.88}$$

or

$$\mu\left(\Delta_M\left(0\right)\right) < 1. \tag{4.89}$$

4.5.3 Examples

In this subsection, we illustrate the proposed passivity-based interaction measures (PB-IM and SB-IM) and compare them to the small gain-based interaction measure.

High-purity distillation column

Consider a binary high-purity distillation column. As shown in Figure 4.9, the flow rates of distillate (D) and boilup (V) are manipulated to control the top and bottom composition y_D and x_B. The linear model of the D–V configuration of the distillation column is given as [87]

$$\begin{bmatrix} y_D\left(s\right) \\ x_B\left(s\right) \end{bmatrix} = G\left(s\right) \begin{bmatrix} D\left(s\right) \\ V\left(s\right) \end{bmatrix}, \tag{4.90}$$

where

$$G\left(s\right) = \frac{1}{1 + 75s} \begin{bmatrix} -0.878\frac{1-0.2s}{1+0.2s} & 0.014 \\ -1.082\frac{1-0.2s}{1+0.2s} & -0.014\frac{1-0.2s}{1+0.2s} \end{bmatrix}.$$

Here we model the interactions as multiplicative uncertainty. Since there are RHP zeros in the diagonal elements of $G\left(s\right)$, $\Delta_M\left(s\right)$ will have unstable poles. As suggested in [87], the simplest way to get around this problem is to neglect the RHP zeros in the diagonal model used in decentralized control and treat the RHP zeros as uncertainty. Therefore,

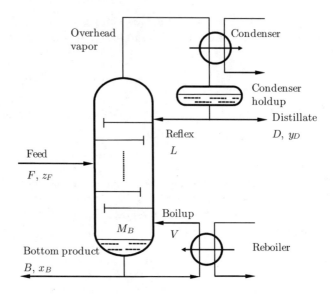

Fig. 4.9. High-purity distillation column

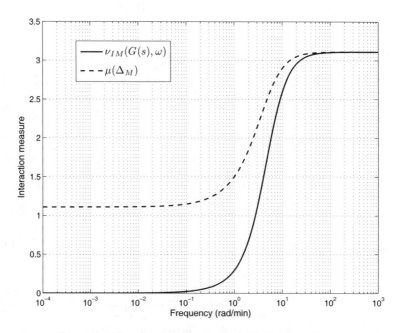

Fig. 4.10. Interaction measures of distillation column

$$G_d(s) = \frac{1}{1 + 75s} \begin{bmatrix} -0.878 & 0 \\ 0 & -0.014 \end{bmatrix}. \tag{4.91}$$

This is actually not very conservative because the RHP zeros limit the achievable performance anyway. The PB-IM and SSV-IM are shown in Figure 4.10. The value of SB-IM depends on the upper sector bound b. For $b = 1000$, $\nu_{sIM}(G(s), \omega, b) \approx \nu_{IM}(G(s), \omega)$. The PB-IM is significantly smaller than the SSV-IM. Particularly, at steady state, $\nu_{IM}(G(s), 0) = 2 \times 10^{-5}$, but $\mu(\Delta_M(0)) = 1.1$. According to (4.88) and (4.89), a decentralized controller that satisfies the passivity-based condition given in Theorem 4.17 can be found to achieve offset-free control, but no controller with integral action satisfies the SSV-IM based condition.

Boiler furnace

The system under consideration is a furnace operating with four burners and four heating coils, as shown in Figure 4.11. Each set of burners is directed at one of the sets of heating coils, but heat naturally spills over to adjacent coils. The coil temperatures T_1, T_2, T_3 and T_4 are controlled by manipulating the gas flow rates F_1, F_2, F_3 and F_4 of the burners, respectively.

The transfer function of the plant is given as follows [99]:

$$\begin{bmatrix} T_1(s) \\ T_2(s) \\ T_3(s) \\ T_4(s) \end{bmatrix} = G(s) \begin{bmatrix} F_1(s) \\ F_2(s) \\ F_3(s) \\ F_4(s) \end{bmatrix}, \tag{4.92}$$

where

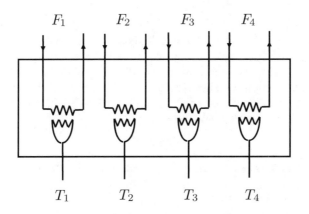

Fig. 4.11. Boiler furnace

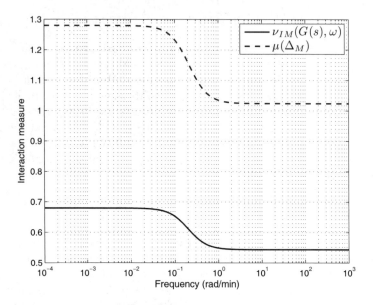

Fig. 4.12. Interaction measures of boiler furnace

$$
G\left(s\right) = \begin{bmatrix}
\frac{1.0}{1+4s} & \frac{0.7}{1+5s} & \frac{0.3}{1+5s} & \frac{0.2}{1+5s} \\
\frac{0.6}{1+5s} & \frac{1.0}{1+4s} & \frac{0.4}{1+5s} & \frac{0.35}{1+5s} \\
\frac{0.35}{1+5s} & \frac{0.4}{1+5s} & \frac{1.0}{1+4s} & \frac{0.6}{1+5s} \\
\frac{0.2}{1+5s} & \frac{0.3}{1+5s} & \frac{0.7}{1+5s} & \frac{1.0}{1+4s}
\end{bmatrix}.
$$

Here the pairing of 1-1/2-2/3-3/4-4 is chosen. Obviously, there are large interactions between different control channels. However, due to the symmetrical
structure of $G(s)$, any PI decentralized controllers with positive parameters
can stabilize the plant [75]. The PB-IM and SSV-IM are shown in Figure 4.12,
from which it is clear that the PB-IM is much less conservative than the SSV-
IM. At steady state, the SSV-IM is 1.28, indicating that the interaction is too
large to use any decentralized controller with an integrator. The PB-IM at
steady state is 0.68, showing that the condition specified in (4.88) is satisfied.

In the next section, we show the decentralized control design for the boiler
furnace example. A controller for the high-purity distillation column can be
obtained by following the same steps.

4.6 Decentralized Control Based on Passivity

4.6.1 Problem Formulation

Decentralized control design can be based on the interaction measures described in the previous section. Because the interactions are modelled as un-

certainties, the decentralized control problem can be solved in the framework of robust control. The decentralized control synthesis problem can be described as follows:

Problem 4.20. Given a multivariable LTI process with a transfer function $G(s) \in \mathbb{C}^{m \times m}$, design a decentralized controller $C(s) = \text{diag}\{c_i(s)\}$ ($i = 1, \ldots, m$), such that

1. The closed-loop system consisting of $G(s)$ and $C(s)$ is stable. This can be achieved by designing a controller $C(s)$ that satisfies the decentralized stability conditions in Theorem 4.17 or Theorem 4.19.
2. The following control performance is achieved:

$$\|W_{SC}S\|_\infty \leq 1, \tag{4.93}$$

where $W_{SC}(s)$ is a user-specified weighting function.

The decentralized design for the closed-loop system performance given in (4.93) can be very hard. One alternative approach is to design each control loop to meet the following specification:

$$\|W_S S_d\|_\infty \leq 1, \tag{4.94}$$

where $S_d(s) = [I + G_d(s)C(s)]^{-1}$ and

$$W_S(s) = \text{diag}\{w_{S1}(s), \ldots, w_{Si}(s), \ldots, w_{Sm}(s)\}, \tag{4.95}$$

i.e.,

$$\left\| w_{Si} [1 + g_{ii}c_i]^{-1} \right\|_\infty \leq 1, \quad \text{for } i = 1, \cdots, m, \tag{4.96}$$

and then check whether (4.93) is satisfied. If this condition is not satisfied, $W_S(s)$ is adjusted and control loops are redesigned until the required closed-loop performance is obtained.

Now the decentralized control problem becomes exactly the robust control design for each loop, subject to the passivity conditions given in Theorems 4.17 and 4.19. If SB-IM is used to characterize the interactions, the PBRC synthesis approaches presented in Section 3.3 can be directly used to design individual control loops.

Alternatively, based on PB-IM, the frequency range in which $\Delta_M(s)$ is "near-passive" can be obtained. The interactions at higher frequencies can be quantified by $\bar{\sigma}(\Delta_M(j\omega))$. Then, the control design presented in Section 3.4 can be implemented to find individual controller loops.

4.6.2 Decentralized Control of Boiler Furnace

Here we design a decentralized control system for the boiler furnace described in Section 4.5.3. This control design example was first published in [11]. A decentralized controller $C(s) = \text{diag}\{c_1(s), c_2(s), c_3(s), c_4(s)\}$ is to be designed such that it stabilizes the full model and achieves zero steady-state tracking error with a settling time less than 25 minutes.

The full model is not diagonal dominant with large interactions. The SSV-IM at steady state is $\mu[\Delta_M(0)] = 1.28$, indicating that it is not possible to design a decentralized controller to satisfy the small gain-based condition given in (4.89). The SB-IM $\nu_{sIM}[G(s), \omega, 1000]$ is almost identical to PB-IM as shown in Figure 4.12, with $\nu_{sIM}[G(s), 0, 1000] = 0.68$. We use the robust control synthesis approach described in Section 3.3.4 to design each control loop.

By curve fitting, the SB-IM is found to be bounded by

$$\nu_{sIM}(G(s), \omega, 1000) \le -\nu_{S-}\left(\frac{s+3.5}{0.01s+5}, \omega, 1000\right). \tag{4.97}$$

Therefore, each closed-loop should satisfy the following condition:

$$\frac{m_{M,i}(s)}{1 - w_2(s)m_{M,i}(s)} + \frac{1}{2000} \quad \text{is ESPR,} \quad \text{for } i = 1, \ldots, 4, \tag{4.98}$$

where $m_{M,i}(s)$ is given in (4.83) and

$$w_2(s) = \frac{s+3.5}{0.01s+5}. \tag{4.99}$$

To satisfy the closed-loop performance specification, the weighting function for each single loop performance is chosen as

$$w_1(s) = \frac{1}{2s}. \tag{4.100}$$

The weighting function to penalize controller gain is $w_3 = 10^{-5}$. The decentralized controller is then obtained by following Procedure 3.16:

$$c_i(s) = \frac{2.46 \times 10^6 s^2 + 1.23 \times 10^9 s + 3.08 \times 10^8}{s^3 + 5.61 \times 10^5 s^2 + 3.41s}, \tag{4.101}$$

for $i = 1 \ldots 4$.

The sensitivity function of the closed-loop system, consisting of the full model and the decentralized controller, $S(s) = [I + G(s)C(s)]^{-1}$, is shown in Figure 4.13. The responses of T_1, T_2, T_3 and T_4 to the step change of F_1 are given in Figure 4.14 (the responses to step changes in F_2, F_3 and F_4 are similar), from which it can be seen that the design specifications are satisfied.

4.7 Summary

In this chapter, we have discussed how the concept of passivity can be used in decentralized control. It is worth pointing out that passivity-based interaction analysis shows the *destabilizing effects* of process interactions rather than

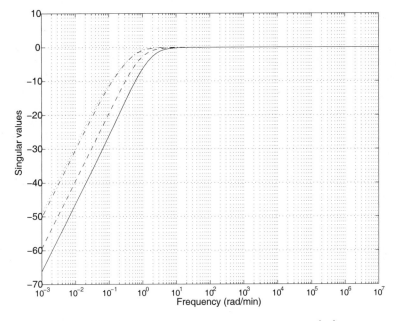

Fig. 4.13. Furnace control: sensitivity function [11]

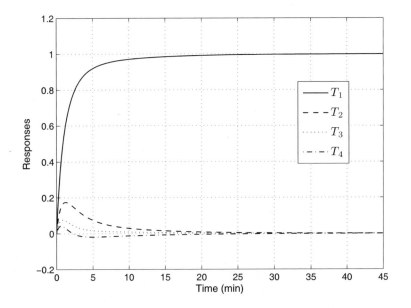

Fig. 4.14. Furnace control: step change responses [11]

the magnitude of interactions. Both the steady-state (DIC and BDIC) and the dynamic interaction analysis tools can be used in determining suitable manipulated/controlled variable pairing schemes. By modelling process interactions as uncertainties, the passivity-based robust control synthesis approaches developed in Chapter 3 can be used to design decentralized controllers.

5

Passivity-based Fault-tolerant Control

We have seen in Chapter 4 that the concept of passivity can be used to develop decentralized control approaches. A passivity-based decentralized controller, if properly designed, can also achieve fault tolerance. For example, for a given multivariable linear strictly passive process, a decentralized passive controller can maintain the stability of a closed-loop control system when one or more controller loops fail simply because the decentralized passive controller remains passive when one or more elements are detuned or taken out of service. Motivated by this observation, a passivity-based decentralized fault-tolerant control framework was developed [16, 17, 148, 149]. In this chapter, we will present the main results of this approach, including fault-tolerant control systems design with zero or minimum control loop redundancy.

5.1 Introduction

In process control applications, failures of control components such as actuators, sensors or controllers are often encountered. A burned-out thermocouple, a broken transducer or a stuck valve are typical control fault events. These problems not only degrade the performance of the control system, but also may induce instability, which could cause serious safety problems. With the increasing reliance on automatic control systems, fault-tolerant control (FTC) becomes an important issue in the process industries. A fault-tolerant control system often consists of a fault detection and diagnosis (FDD) subsystem and the so-called fault accommodation subsystem, which is the controller that achieves stability and/or control performance objective when faults happen [20].

At present, most existing fault-tolerant control systems are built with redundant controllers (*e.g.*, [140]). Two different approaches are commonly used in control practice: active and passive redundancies. In the active redundancy scheme, a backup controller is activated once a fault of the main control loop is detected online [111]. This approach has some obvious drawbacks. First, control loop faults may not be detected swiftly and accurately. Second, the fault detection system itself could be a possible source of failures. In the passive redundancy scheme, the FTC problem is formulated as a reliable stabilization problem [131]. The process is connected to both the main controller and the backup controller and can be stabilized by using either or both controller(s). Therefore, system stability is guaranteed in the presence of control loop failures without requiring fault detection and subsequent controller switching. Both above approaches require a significant number of redundant control components, which sometimes may lead to an unacceptable level of capital and maintenance costs.

Redundancy-free fault-tolerant control has been studied recently. One approach is to detect and diagnose system faults online and update control laws accordingly to stabilize the faulty system [154]. For example, when sensor/actuator failure occurs, the controller parameters are adjusted or even different controllers, which are designed based on usable sensors and actuators, are employed (e.g., [83, 142]). However, this method could be very complicated to implement because it requires preprogramming multiple control algorithms for every possible failure scenario. Another approach is to design fixed decentralized controllers that are inherently fault-tolerant so that no FDD systems are required.

In this chapter, we focus mainly on the issue of fault accommodation, including the analysis for stability under control component failure and design methods for fault-tolerant controllers. For a stable linear process, it is possible to design a decentralized controller such that the closed-loop system remains stable when one or more subloops are partially or fully switched off by sensor or actuator faults. The stability under this circumstance is called decentralized unconditional stability (DUS) [26] or decentralized detunability [58]. In later sections, we will present a framework for decentralized unconditional stabilizing control on the basis of passivity and its extensions. We show how this framework can be combined with active fault-tolerant control approaches to deliver reliable and cost-effective solutions.

5.2 Representation of Sensor/Actuator Faults

The faults encountered in process control practice can often be classified as plant faults, sensor faults and/or actuator faults [20]. This can be illustrated by the control system depicted in Figure 5.1. The process has m inputs and m outputs and can be represented by a transfer function $G(s) \in \mathbb{C}^{m \times m}$, where $u_a \in \mathbb{R}^m$ is the actuator output vector and $y \in \mathbb{R}^m$ is the process output vector. The outputs from the controller and sensor are $u \in \mathbb{R}^m$ and $y_m \in \mathbb{R}^m$, respectively. Assume that the dynamics of sensors and actuators are negligible. The plant faults refer to the hardware failure of the plant itself, leading to changes in process system dynamics (i.e., in $G(s)$).

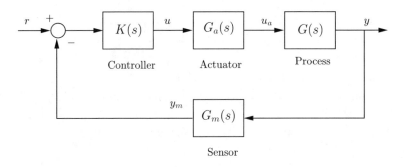

Fig. 5.1. Decentralized feedback control system

In this section, we focus on sensor and actuator faults. Actuator and sensor faults can be represented by the following models [17]:

$$u_a = E_a u + f_a,$$
$$y_m = E_m y + f_m, \tag{5.1}$$

where E_a, $E_m \in \mathbb{R}^{m \times m}$ are actuator and sensor fault matrices with the following diagonal structures:

$$E_a = \begin{bmatrix} \varepsilon_1^a & 0 & \cdots & 0 \\ 0 & \varepsilon_2^a & \cdots & 0 \\ \vdots & \vdots & \ddots & \vdots \\ 0 & 0 & \cdots & \varepsilon_m^a \end{bmatrix}, \quad E_m = \begin{bmatrix} \varepsilon_1^m & 0 & \cdots & 0 \\ 0 & \varepsilon_2^m & \cdots & 0 \\ \vdots & \vdots & \ddots & \vdots \\ 0 & 0 & \cdots & \varepsilon_m^m \end{bmatrix}, \tag{5.2}$$

where $0 \le \varepsilon_i^a, \varepsilon_i^m \le 1$ $(i = 1, 2, \ldots, m)$. Vectors $f_a = [f_1^a, \ldots, f_i^a, \ldots, f_m^a]^T$ and $f_m = [f_1^m, \ldots, f_i^m, \ldots, f_m^m]^T$ represent the constant components of actuator and sensor outputs when they fail. This model addresses the following typical fault scenarios (assuming that the ith channel of the control system fails):

1. Sensor outage: $\varepsilon_i^m = 0$, $f_i^m = 0$;
2. Controller/actuator outage: $\varepsilon_i^a = 0$, $f_i^a = 0$;
3. Sensor partially functioning: $0 \le \varepsilon_i^m \le 1$;
4. Actuator partially functioning: $0 \le \varepsilon_i^a \le 1$;
5. Frozen sensor output: $\varepsilon_i^m = 0$, $f_i^m =$ constant output from the ith sensor;
6. Frozen controller output and/or actuator stickiness: $\varepsilon_i^a = 0$, $f_i^a =$ constant output from the ith controller/actuator.

For the linear feedback system under consideration (as shown in Figure 5.2), constant vectors f_a and f_m do not affect the closed-loop stability. Consequently, for stability analysis under the control failure scenarios listed above, only the effects of actuator and sensor fault matrices E_a and E_m need to be considered. Due to their diagonal structures, controller $K(s)$ and fault

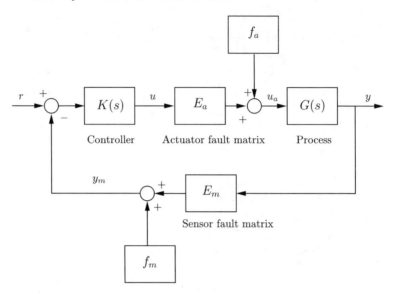

Fig. 5.2. Representation of sensor and actuator failures [17]

matrices E_a and E_m are permutable, and thus sensor/actuator failures can be treated as detuning of the decentralized controller $K(s)$. Therefore, fault tolerance can be achieved by a decentralized unconditionally stabilizing controller $K(s)$, which ensures closed-loop stability when one or more of its outputs are arbitrarily detuned or switched off.

5.3 Decentralized Unconditional Stability Condition

In this section, we discuss under what condition a control system can achieve decentralized unconditional stability (DUS) and the implications of this condition. This section is based mainly on the results developed in [149] by the authors and their collaborator.

5.3.1 Passivity-based DUS Condition

Clearly, if a linear process is strictly passive (*i.e.*, stable and strictly input passive), any passive decentralized controller can achieve DUS. If the process is not strictly passive, the controller needs to obey the following additional conditions:

Theorem 5.1 (Passivity-based DUS condition[149]). *For the interconnected system comprised of a stable process $G(s) \in \mathbb{C}^{m \times m}$ and a decentralized*

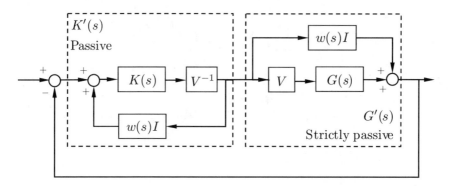

Fig. 5.3. DUS analysis [149]

controller $K(s) = \text{diag}\{k_i(s)\}$, $(i = 1, \ldots, m)$, as shown in Figure 5.3, if a passive transfer function $w(s)$ can be found such that

$$\nu_-\left(w(s), \omega\right) < -\nu_-\left(G^+(s), \omega\right), \tag{5.3}$$

then the closed-loop system will be decentralized unconditionally stable if for any loop i $(i = 1, \ldots, m)$

$$k_i'(s) = k_i^+(s)\left[1 - w(s)k_i^+(s)\right]^{-1} \text{ is passive,} \tag{5.4}$$

where

$$V = \text{diag}\{V_{ii}\}, \quad i = 1, \ldots, m \tag{5.5}$$

is a diagonal matrix with either 1 or -1 along the diagonal. The signs of elements of V are determined such that the diagonal elements of

$$G^+(s) = G(s)V \tag{5.6}$$

are positive at steady state. The SISO system $k_i^+(s)$ is the ith element of the diagonal system

$$K^+(s) = V^{-1}K(s) = VK(s). \tag{5.7}$$

Proof. The stability of the closed-loop system in Figure 5.1 (assume that $G_a = I$ and $G_m = I$) is equivalent to that system in Figure 5.3 by loop shifting, where

$$G'(s) = G(s)V + w(s)I, \tag{5.8}$$

$$K'(s) = VK(s)\left[I - w(s)VK(s)\right]^{-1}. \tag{5.9}$$

Since $G'(s)$ is strictly passive, from the Passivity Theorem, the closed-loop system in Figure 5.3 is stable if $K'(s)$ is passive (*i.e.*, $k_i'(s)$ is passive for $i = 1, \ldots, m$). When loop i $(i = 1, \ldots, m)$ of the multiloop controller $k_i(s)$ is

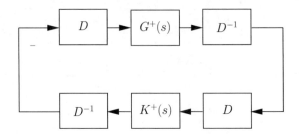

Fig. 5.4. Diagonal scaling of passivity index [149]

arbitrarily detuned to $\varepsilon_i k_i(s)$ $(0 \le \varepsilon_i \le 1)$, the corresponding $k'_i(s)$ (denoted as $k'_{i,d}(s)$) is given by the following equations:

$$k'_{i,d}(s) = \frac{V_{ii}\varepsilon_i k_i(s)}{1 - w(s)V_{ii}\varepsilon_i k_i(s)}, \tag{5.10}$$

as

$$k_i(s) = \frac{k'_i(s)}{V_{ii}\left(1 + w(s)k'_i(s)\right)}. \tag{5.11}$$

Then,

$$k'_{i,d}(s) = \frac{\frac{\varepsilon_i k'_i(s)}{1+w(s)k'_i(s)}}{1 - w(s)\frac{\varepsilon_i k'_i(s)}{1+w(s)k'_i(s)}} = \frac{\varepsilon_i k'_i(s)}{1 + (1 - \varepsilon_i)w(s)k'_i(s)}. \tag{5.12}$$

If $k'_i(s)$ is passive for $i = 1, \ldots, m$, the right-hand side of (5.12) can be interpreted as the negative feedback system of two passive subsystems $\varepsilon_i k'_i(s)$ and $\frac{1-\varepsilon_i}{\varepsilon_i}w(s)$. This leads to the conclusion that $k'_{i,d}(s)$ is passive for all $\varepsilon_i \in [0, 1]$. Therefore, decentralized unconditional stability of the closed-loop system can be achieved.

5.3.2 Diagonal Scaling

Because the controller $K(s)$ is decentralized, the DUS condition given in Theorem 5.1 can be made less conservative by diagonally scaling the passivity index of the process. This is very similar to the diagonal scaling treatment used in the passivity-based interaction measure described in Section 4.5.2. The only difference is that here we scale the process model rather than the uncertainty system.

Assume that D is a diagonal and nonsingular matrix. The closed-loop system in Figure 5.1 (with perfect sensors and actuators) is stable if and only if the system in Figure 5.4 is stable (where $G^+(s)$ and $K^+(s)$ are defined in (5.6) and (5.7)). Note that for any diagonal system $K^+(s)$, $K^+(s) = D^{-1}K^+(s)D$.

However, the passivity index of $D^{-1}G^+(s)D$ can be significantly reduced by choosing an appropriate frequency-dependent D matrix. After replacing $\tilde{\Delta}(s)$ with $G^+(s)$, the scaling matrix D can be found by solving Problem 4.14 following the numerical method given in Problem 4.15. The diagonally scaled passivity index can be defined as

$$\nu_D\left(G^+(s),\omega\right) \triangleq -\frac{1}{2}\lambda\left\{D^{-1}(\omega)G^+(j\omega)D(\omega) + D(\omega)\left[G^+(j\omega)\right]^* D^{-1}(\omega)\right\}. \tag{5.13}$$

Therefore, (5.3) in Theorem 5.1 can be replaced by the following condition:

$$\nu_-\left(w(s),\omega\right) < -\nu_D\left(G^+(s),\omega\right). \tag{5.14}$$

Now let us study the effects of diagonal scaling. Given a stable process transfer matrix with signs adjusted, $G^+(s) \in \mathbb{C}^{m\times m}$, diagonal scaling of the process at frequency ω gives:

$$
\begin{aligned}
&D(\omega)^{-1}G^+(j\omega)D(\omega) \\[2mm]
&= \begin{bmatrix} d_1^{-1} & 0 & \cdots & 0 \\ 0 & d_2^{-1} & \cdots & 0 \\ \vdots & \vdots & \ddots & \vdots \\ 0 & 0 & \cdots & d_m^{-1} \end{bmatrix} \begin{bmatrix} G_{11}^+ & G_{12}^+ & \cdots & G_{1m}^+ \\ G_{21}^+ & G_{22}^+ & \cdots & G_{2m}^+ \\ \vdots & \vdots & \ddots & \vdots \\ G_{m1}^+ & G_{m2}^+ & \cdots & G_{mm}^+ \end{bmatrix} \begin{bmatrix} d_1 & 0 & \cdots & 0 \\ 0 & d_2 & \cdots & 0 \\ \vdots & \vdots & \ddots & \vdots \\ 0 & 0 & \cdots & d_m \end{bmatrix} \\[2mm]
&= \begin{bmatrix} G_{11}^+ & d_1^{-1}G_{12}^+ d_2 & \cdots & d_1^{-1}G_{1m}^+ d_m \\ d_2^{-1}G_{21}^+ d_1 & G_{22}^+ & \cdots & d_2^{-1}G_{2m}^+ d_m \\ \vdots & \vdots & \ddots & \vdots \\ d_m^{-1}G_{m1}^+ d_1 & d_m^{-1}G_{m2}^+ d_2 & \cdots & G_{mm}^+ \end{bmatrix}.
\end{aligned} \tag{5.15}
$$

Clearly, diagonal scaling affects only the off-diagonal elements of $G^+(j\omega)$ to make $G^+(s)$ more passive. The effectiveness of diagonal scaling can be seen from the distillation column example in Section 5.4.1.

5.3.3 Achievable Control Performance

If $\nu_D(G^+(s),\omega) > 0$, the decentralized unconditional stability condition given in (5.3) implies that

1. $k_i^+(s)$ is passive.
2. The frequency response of each controller loop k_i at frequency ω is confined in the disc centered at $(1/\left[2\nu_D(G^+(s),\omega)\right], 0)$ with a radius of $1/\left|2\nu_D(G^+(s),\omega)\right|$ (as shown in Figure 5.5), *i.e.*,

$$\left|k_i^+(j\omega) - \frac{1}{2\nu_D(G^+(s),\omega)}\right| \leq \left|\frac{1}{2\nu_D(G^+(s),\omega)}\right|, \quad \forall\, i,\omega. \tag{5.16}$$

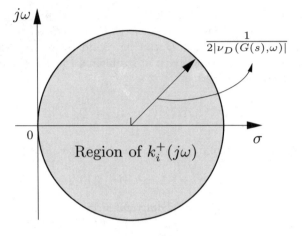

Fig. 5.5. Frequency response of DUS controller

The size of the disc changes with frequency, but the disc is tangential to the imaginary axis. Inequality 5.16 implies the limitation on the achievable performance of the passivity-based controller: At a particular frequency, the larger the passivity index $\nu_D(G^+(s),\omega)$, the smaller the controller amplitude ratio allowed.

The above condition is actually the extension of the DIC condition given in Theorem 4.4. If the process $G(s)$ is stable with $\nu_D(G^+(s),0) \leq 0$, then the process is DIC. In this case, there exists a frequency band $[0,\omega_b]$, in which $\nu_D(G^+(s),\omega) \leq 0, \forall \omega \in [0,\omega_b]$. Assuming no constraints on controller gain, arbitrarily large amplitude ratios of the controller are possible in this frequency band. The larger the upper bound ω_b, the larger bandwidth the passive decentralized controller can have, and the faster the response that can be achieved.

5.3.4 Pairing for Dynamic Performance

The analysis of performance achievable by a DUS controller described in Section 5.3.3 can be used to determine suitable pairing schemes for DUS control design. Different pairing schemes result in different transfer functions $G(s)$, which normally have different passivity indices. Because the diagonally scaled passivity index implies a constraint on the achievable performance of a passive DUS controller, a pairing scheme should be chosen such that the resulting $G(s)$ has small $\nu_D(G^+(s),\omega)$ at all frequencies concerned. The pairing procedure for DUS control can be described as follows:

Procedure 5.2 (Pairing for DUS control [15])

1. Determine the transfer function $G(s)$ for each possible pairing scheme.

2. *Find the sign matrix V and obtain $G^+(s)$ such that $G_{ii}^+(0) > 0$ ($i = 1, \ldots, m$).*
3. *Screen out the non-DIC pairing schemes by using the necessary DIC condition given in Theorem 4.3.*
4. *Calculate the diagonally scaled passivity indices $\nu_D(G^+(s), \omega)$ at a number of frequency points.*
5. *Compare the passivity index profiles of different pairings. The best pairing should correspond to the one with the largest frequency bandwidth ω_b such that $\nu_D(G^+(s), \omega) \leq 0$ for any $\omega \in [0, \omega_b]$. This pairing scheme would allow using controllers with integral action and the fastest dynamic response.*

Examples

Now we illustrate how to use the above pairing procedure for DUS control with two examples.

Example 5.3 (Distillation column [15]). Consider the distillation column described by the following 3×3 transfer matrix [79]:

$$G(s) = \begin{bmatrix} \dfrac{-1.986e^{-0.71s}}{66.67s+1} & \dfrac{5.24e^{-60s}}{400s+1} & \dfrac{5.984e^{-2.24s}}{14.29s+1} \\ \dfrac{0.0204e^{-4.199s}}{5s+1} & \dfrac{-0.33e^{-1.883s}}{3.904s+1} & \dfrac{2.38e^{-1.143s}}{10s+1} \\ \dfrac{0.374e^{-7.75s}}{22.22s+1} & \dfrac{-11.3e^{-14.78s}}{33.66s+1} & \dfrac{-9.881e^{-1.59s}}{11.35s+1} \end{bmatrix}. \tag{5.17}$$

The pairing schemes 1-2/2-3/3-1, 1-2/2-1/3-3 and 1-3/2-2/3-1 are not desirable because they do not satisfy the DIC condition, as $\nu_D(G^+(s), 0) > 0$. As a result, only three pairing schemes remain, *i.e.*, 1-1/2-2/3-3, 1-3/2-1/3-2 and 1-1/2-3/3-2. The diagonally scaled passivity indices of the resulting transfer function were calculated and plotted in Figure 5.6 for each pairing scheme. According to the proposed pairing rule, the pairing 1-1/2-3/3-2, which has the largest passivity bandwidth, is preferred. This is consistent with the conclusion from the generalized dynamic relative gain (GDRG) method [60] because the pairing scheme of 1-1/2-3/3-2 also has the smallest loop interactions measured by the so-called total interaction potential.

Note that suitable pairing schemes for DUS control may not always lead to small loop interactions. This is clearly demonstrated in the following example:

Example 5.4 ([15]). Consider another 2×2 process with the following transfer function [60]:

$$G(s) = \begin{bmatrix} \dfrac{-2e^{-s}}{10s+1} & \dfrac{1.5e^{-s}}{s+1} \\ \dfrac{1.5e^{-s}}{s+1} & \dfrac{-2e^{-s}}{10s+1} \end{bmatrix}. \tag{5.18}$$

The diagonally scaled passivity indices of the process with two different pairing schemes are shown in Figure 5.7. Off-diagonal pairing has smaller

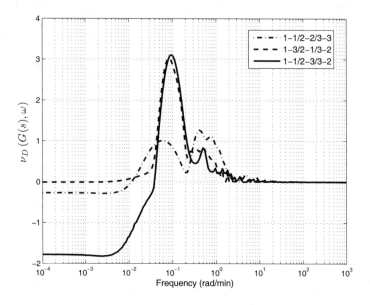

Fig. 5.6. Example 5.3: Scaled passivity index

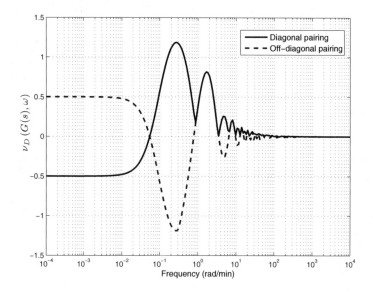

Fig. 5.7. Example 5.4: Scaled passivity index

loop interactions (reflected by a smaller total interaction potential) than the diagonal pairing and thus is preferred according to the GDRG pairing criterion [60]. However, the DUS pairing rule prefers the diagonal pairing scheme. Off-diagonal pairing has a positive ν_D index at steady state and thus is not DIC. While diagonal pairing does lead to larger dynamic interaction which is evident by the GDRG analysis, the destabilizing effect of the interaction is insignificant. This was confirmed by a simulation study which showed that a DUS controller can be used to stabilize the above process. The simulation study also confirmed that the diagonal pairing scheme allowed using controllers that achieve high control performance [15].

5.4 Fault-tolerant Control Design for Stable Processes

In this section, we show how to design a fault-tolerant control system for a stable linear process on the basis of the DUS condition given in Section 5.3.

5.4.1 Fault-tolerant PI Control

We first look at the fault-tolerant multiloop PI controller tuning problem. This is perhaps an interesting topic to process control engineers because PI controllers are widely used in process industries. Since multiloop PI controllers with positive gains are passive, they can be further tuned to satisfy passivity-based DUS conditions. In this section, an optimization-based tuning approach that achieves both DUS and certain performance specification is presented.

Problem formulation

For multiloop PI controllers, Theorem 5.1 is reduced to the following condition:

Proposition 5.5 ([14]). *Given a stable LTI MIMO process with its transfer function $G(s) \in \mathbb{C}^{m \times m}$, any multiloop PI controller*

$$K(s) = \operatorname{diag}\{k_i(s)\} = \operatorname{diag}\left\{k_{c,i}\left(1 + \frac{1}{\tau_{I,i}s}\right)\right\}, \quad \forall \, i = 1, \ldots, m, \quad (5.19)$$

with the following parameter relation will unconditionally stabilize the closed-loop system:

$$\tau_{I,i}^2 \geq \frac{k_{c,i}^+ \nu_D(G^+(s), \omega)}{\left[1 - k_{c,i}^+ \nu_D(G^+(s), \omega)\right]\omega^2} \quad \forall \, \omega, \, i, \quad (5.20)$$

where $G^+(s) = G(s)V$ and $k_{c,i}^+ = k_{c,i}V_{ii} \geq 0$.

There are different ways to represent the performance specification for multiloop control. A simple approach is to use the sensitivity function of each loop $S_i(s)$ $(i = 1, \ldots, m)$ with a weighting function $w_i(s)$ that penalizes control error at low frequencies. Define γ_i as a scalar decision variable for loop i. The controller loops can be designed to maximize γ_i subject to $|\gamma_i S_i(j\omega) w_i(j\omega)| < 1$ for all ω and $i = 1, \ldots, m$. The PI tuning problem is converted into the following optimization problem:

Problem 5.6 (DUS PI controller tuning [14]). For the ith controller $k_i(s)$,

$$\min_{k_{c,i}^+, \tau_{I,i}} \{-\gamma_i\}, \tag{5.21}$$

subject to

$$\left| \frac{w_i(j\omega)\gamma_i}{1 + G_{ii}^+(j\omega)k_{c,i}^+\left[1 + 1/(j\omega\tau_{I,i})\right]} \right| < 1, \tag{5.22}$$

and (5.20) is satisfied $\forall\, \omega \in \mathbb{R}$, $i = 1, \ldots, m$.

One of the possible choices of the weighting function is $w_i(s) = 1/s$ so that offset-free control can be achieved. A larger γ_i in (5.22) implies a larger bandwidth of the control system and therefore a faster response and smaller disturbance impact. The above optimization problem can also be solved using any nonlinear optimization tool, such as the MATLAB® *Optimization Toolbox*.

PI controller tuning procedure

For a given stable process $G(s)$, a multiloop fault-tolerant PI controller can be obtained by solving Problem 5.6 using the design procedure as follows.

Procedure 5.7 (DUS PI controller tuning)   *1. Determine the pairing scheme for controlled and manipulated variables according to Procedure 5.2. If none of the pairing schemes is not DIC, then it is not possible to design a DUS PI controller.*

2. For each subsystem $G_{ii}^+(s)$ $(i = 1, \ldots, m)$, solve Problem 5.6 for the PI controller parameters $k_{c,i}^+$ and $\tau_{I,i}$.

3. Adjust the signs of the final subcontroller gain $k_{c,i} = V_{ii}k_{c,i}^+$ to obtain the final multiloop controller

$$K(s) = \mathrm{diag}\left\{ k_{c,i}\left(1 + \frac{1}{\tau_{I,i}s}\right) \right\}, \quad i = 1, \ldots, m.$$

DUS PI controllers can also be obtained based on existing tuning methods. For example, we can use the Ziegler-Nichols method to find the initial PI controller parameters $k_{ZN,i}$ and $\tau_{ZN,i}$. These parameters can then be tuned independently to satisfy the DUS condition given in (5.20). This tuning approach can be formulated as follows:

Problem 5.8 (DUS PI control tuning based on Ziegler-Nichols method [148]). For the ith controller,

$$\min_{F_{1i}, F_{2i}} F_{1i}^2 + F_{2i}^2 \tag{5.23}$$

such that

$$F_{2i}^2 \tau_{ZN,i}^2 \geq \frac{k_{ZN,i} \nu_s(G^+(s), \omega)}{[F_{1i} - k_{ZN,i} \nu_s(G^+(s), \omega)] \omega^2}, \quad \forall \, \omega, \, i. \tag{5.24}$$

The final PI controller settings are

$$k_{c,i} = \frac{V_{ii} k_{ZN,i}}{F_{1i}}, \quad \tau_{I,i} = \tau_{ZN,i} F_{2i}. \tag{5.25}$$

Example of DUS PI control design

Here we show an example of multiloop DUS PI control tuning for a distillation column [79]. This example was first published in [149]. The process has the following transfer function:

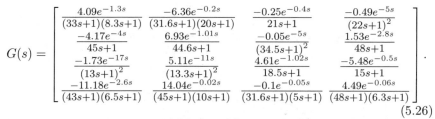

$$(5.26)$$

This process is stable and $G^+(s) = G(s)$ because all diagonal elements of $G(0)$ are positive. For the given pairing scheme 1-1/2-2/3-3/4-4, the original $(\nu_-(G^+(s), \omega))$ and diagonally scaled $(\nu_D(G^+(s), \omega))$ passivity indices are plotted in Figure 5.8, from which it can be seen that the scaled index is significantly smaller. It is observed that the process with this pairing scheme is DIC as $\nu_D(G^+(s), 0) < 0$. The sensitivity weighting function $w_i(s) = \frac{1}{s}$ was chosen for each control loop. The DUS PI controller was designed following Procedure 5.7. For comparison purposes, PI controller parameters were also calculated using the biggest log modulus tuning (BLT) method [79]. In the BLT design, two different "biggest log modulus" values, $L_{max} = 4$ and $L_{max} = 8$), were used (a smaller L_{max} would result in a more robust control system while a larger L_{max} can lead to better performance). The controller parameters from all design approaches are listed in Table 5.1. Simulation studies were performed for all controllers. The performance of the DUS PI controller and the BLT controllers are compared in Table 5.1 in terms of the integral of time-weighted absolute error (ITAE) of the first 1000 minutes. The step-change responses are exemplified by the response of Loop 1, as shown in Figure 5.9.

From Table 5.1 we see in general that the proposed approach achieves better performance than the BLT method with $L_{max} = 4$. Compared with the

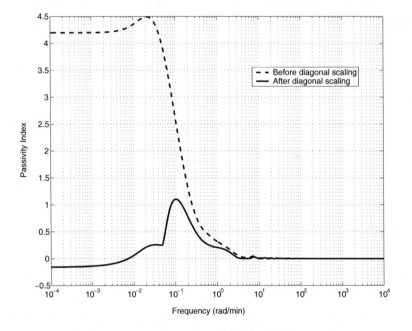

Fig. 5.8. Passivity indices of the distillation column

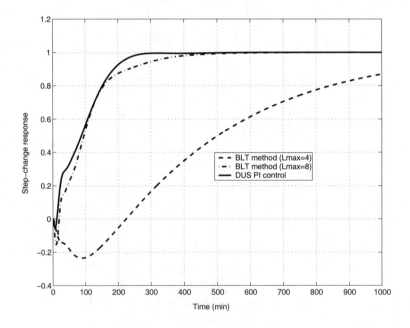

Fig. 5.9. Step-change response of Loop 1 [149]

Table 5.1. Multiloop PI controller settings

Approach	PI controller	Loop 1	Loop 2	Loop 3	Loop 4
BLT method ($L_{max} = 4$)	k_c	0.393	0.495	0.310	0.927
	τ_I	145	31.0	31.0	93.8
	ITAE	1.95×10^5	4400	1.06×10^5	9.12×10^4
BLT method ($L_{max} = 8$)	k_c	0.923	1.160	0.727	2.170
	τ_I	61.7	13.2	13.2	40.0
	ITAE	1.12×10^4	273	4246	4051
DUS PI control	k_c	0.867	0.848	0.772	0.854
	τ_I	47.1	37.7	24.6	40.0
	ITAE	7168	1258	9262	1.13×10^4

BLT controller design with $L_{max} = 8$, the DUS controller performs better in Loop 1 but worse in other loops. This can be explained because the proposed design method is based on the passivity index of the entire process model, which reflects the overall stability constraints resulting from both the process dynamics and loop interactions. The large time delay and the second-order dynamics in $G_{11}(s)$ lead to a large overall passivity index. This large passivity index imposes a constraint on the performance achievable for each loop. From this example, it can be seen that decentralized unconditional stability is a demanding requirement and as such, controllers may have to sacrifice performance to achieve DUS.

5.4.2 Decentralized Fault-tolerant $\mathcal{H}_2$ Control Design

Now we present the design approach of general DUS control systems that achieve $\mathcal{H}_2$ control performance. For a given system with a transfer function $G(s)$, its $\mathcal{H}_2$ norm is defined as follows [156]:

$$\|G\|_2 \triangleq \sup_{\sigma > 0} \left\{ \frac{1}{2\pi} \int_{-\infty}^{\infty} \text{tr} \left[G^* \left(\sigma + j\omega \right) G \left(\sigma + j\omega \right) \right] d\omega \right\}^{\frac{1}{2}}. \tag{5.27}$$

If $G(s)$ is analytic in $\text{Re}(s) > 0$, then

$$\|G\|_2 = \left\{ \frac{1}{2\pi} \int_{-\infty}^{\infty} \text{tr} \left[G^* \left(j\omega \right) G \left(j\omega \right) \right] d\omega \right\}^{\frac{1}{2}}. \tag{5.28}$$

In this design, the $\mathcal{H}_2$-norm of the closed-loop system that represents the control performance is minimized. This subsection is based mainly on the work developed by the authors and their co-worker [16].

The decentralized fault-tolerant $\mathcal{H}_2$ controller design problem can be stated as follows: Given an $m \times m$ LTI stable process with a transfer function $G(s)$, find a decentralized controller $K(s) = \text{diag} \{k_i(s)\}$ $(i = 1, \ldots, m)$ such that the DUS condition is satisfied and an $\mathcal{H}_2$ nominal performance for each control loop is optimized. If a stable and minimum phase weighting function $w(s)$ can be chosen such that $\nu_- (w(s), \omega) < -\nu_D (G^+(s), \omega)$, where $G^+(s)$ is given in (5.6), the controller design problem of designing the ith controller $(i = 1, \ldots, m)$ can be mathematically formulated as follows:

Problem 5.9.

$$\min_{k_i(s)} \{\gamma_i\}, \tag{5.29}$$

subject to

$$k_i'(s) = V_{ii} k_i(s) \left[1 - w(s) V_{ii} k_i(s)\right]^{-1} \text{ is passive, } \forall \ i = 1, \ldots, m, \tag{5.30}$$

and

$$\|T_i (G_{ii}, k_i)\|_2 < \gamma_i, \tag{5.31}$$

where $T_i (G_{ii}(s), k_i(s))$ is the closed-loop transfer function (representing the control performance) whose $\mathcal{H}_2$-norm needs to be minimized. $G_{ii}(s)$ is the ith diagonal element of $G(s)$ and V_{ii} is the ith diagonal element of the sign matrix V given in (5.5).

A typical choice of the closed-loop transfer function is as follows:

$$T_i (G_{ii}(s), k_i(s)) = \begin{bmatrix} w_{ki} k_i(s) \left[I + G_{ii}(s) k_i(s)\right]^{-1} \\ w_{si}(s) \left[I + G_{ii}(s) k_i(s)\right]^{-1} \end{bmatrix}, \tag{5.32}$$

where $w_{si}(s)$ is the weighting function for sensitivity and w_{ki} is a constant weight to penalize the controller gain. This is an independent design approach because each control loop should be designed to satisfy the same condition given in (5.30).

5.4.3 Selecting the Weighting Function $w(s)$

The weighting function $w(s)$ must be stable and minimum phase because $w(s)$ will be absorbed into the final controllers. It is also desirable that this transfer function is simple and of low order. If $G^+(s)$ is DIC, a typical $w(s)$ could be chosen to have the following form:

$$w(s) = \frac{ks(s + a)}{(s + b)(s + c)}, \tag{5.33}$$

where a, b, c and k are positive real parameters to be determined.

If $G^+(s)$ is not DIC, the weighting function $w(s)$ does not possess a zero at $s = 0$. In this case, the following form can be used:

$$w(s) = \frac{k(s + a)(s + b)}{(s + c)(s + d)}. \tag{5.34}$$

With such a weighting function $w(s)$, the final controller $K(s) = \text{diag} \{k_i(s)\}$ $(i = 1, \ldots, m)$ does not have integral action.

If the passivity indices at n_ω frequency points

$$\left[\nu_D \left(G^+(s), \omega_1 \right), \nu_D \left(G^+(s), \omega_2 \right), \cdots, \nu_D \left(G^+(s), \omega_{n_\omega} \right) \right] \tag{5.35}$$

are obtained, the parameters of $w(s)$ can be found by solving the following optimization problem:

Problem 5.10.

$$\min_{a,b,c,k} \sum_{i=1}^{n_\omega} \left[\mathrm{Re}\left(w\left(j\omega_i \right) \right) - \nu_D \left(G^+(s), \omega_i \right) \right]^2, \tag{5.36}$$

subject to

$$\mathrm{Re}\left(w\left(j\omega_i \right) \right) > \nu_D \left(G^+(s), \omega_i \right), \quad \forall\, i = 1, \ldots, n_\omega. \tag{5.37}$$

Like Problem 5.6, the above problem can be solved by using any nonlinear optimization solver, such as the MATLAB® *Optimization Toolbox*.

5.4.4 Control Synthesis

We can simplify the control design task by constructing the DUS controller $K(s) = \mathrm{diag}\{k_i(s)\}$ $(i = 1, \ldots, m)$ indirectly. From Figure 5.3, we can see that

$$K(s) = VK'(s) \left[I + w(s)K'(s) \right]^{-1}. \tag{5.38}$$

Therefore, we can first design the system $K'(s)$, which is required to be passive, and then obtain the final controller using (5.38). The passive controller $K'(s) = \mathrm{diag}\{k_i'(s)\}$ $(i = 1, \ldots, m)$ can be found by solving the following problem:

Problem 5.11. For all $i = 1, \ldots, m$

$$\min_{k_i'(s)} \{\gamma_i\}, \tag{5.39}$$

subject to

$$k_i'(s) \text{ is passive}, \tag{5.40}$$

and

$$\left\| T_i' \left(G_{ii}, k_i' \right) \right\|_2 < \gamma_i. \tag{5.41}$$

The closed-loop transfer function $T_i \left(G_{ii}(s), k_i'(s) \right)$ gives a performance constraint for the ith loop as a function of $k_i'(s)$. For the equivalent performance specification as given in (5.31),

$$T_i \left(G_{ii}(s), k_i'(s) \right) = \begin{bmatrix} w_{ki} V_{ii} k_i'(s) \left[I + G_{ii}(s) V_{ii} k_i'(s) + w(s) k_i'(s) \right]^{-1} \\ w_{si}(s) \left[I + w(s) k_i'(s) \right] \left[I + G_{ii}(s) V_{ii} k_i'(s) + w(s) k_i'(s) \right]^{-1} \end{bmatrix}. \tag{5.42}$$

For a given LTI stable process $G(s) \in \mathbb{C}^{m \times m}$, assume the following state-space representations (for $i = 1, \ldots .m$):

$$G_{ii}^+(s) := (A_{gi}, B_{gi}, C_{gi}, D_{gi}),$$ (5.43)

$$w(s) := (A_p, B_p, C_p, D_p), \text{ and}$$ (5.44)

$$w_{si}(s) := (A_{wi}, B_{wi}, C_{wi}, D_{wi}).$$ (5.45)

The following augmented plant can be constructed for the ith loop:

$$P_i(s) = \begin{bmatrix} P_{11}(s) & P_{12}(s) \\ P_{21}(s) & P_{22}(s) \end{bmatrix} := \begin{cases} \dot{x} = Ax + B_1 w + B_2 u \\ z = C_1 x + D_{11} w + D_{12} u \,, \\ y = C_2 x + D_{21} w + D_{22} u \end{cases}$$ (5.46)

where

$$A = \begin{bmatrix} A_{gi} & 0 & 0 \\ 0 & A_p & 0 \\ -B_{wi} & 0 & A_{wi} \end{bmatrix}, \quad B_1 = \begin{bmatrix} 0 \\ 0 \\ B_{wi} \end{bmatrix}, \quad B_2 = \begin{bmatrix} B_{gi} \\ B_p \\ -B_{wi}D_{gi} \end{bmatrix},$$

$$C_1 = \begin{bmatrix} -D_{wi}C_{gi} & 0 & C_{wi} \\ 0 & 0 & 0 \end{bmatrix}, \quad D_{11} = \begin{bmatrix} D_p \\ 0 \end{bmatrix}, \quad D_{12} = \begin{bmatrix} -D_p D_{gi} \\ w_{ki} \end{bmatrix},$$

$$C_2 = \begin{bmatrix} -C_{gi} & -C_p & 0 \end{bmatrix}, \quad D_{21} = I, \quad D_{22} = -D_{gi} - D_p,$$

such that

$$\mathcal{F}_l \left(P_i(s), k_i'(s) \right) = P_{11}(s) + P_{12}(s)k_i'(s) \left[I - P_{22}(s)k_i'(s) \right]^{-1} P_{21}(s)$$
$$= T_i \left(G_{ii}(s), k_i'(s) \right).$$ (5.47)

Therefore, the problem becomes finding a passive solution to the $\mathcal{H}_2$ control problem of the augmented plant $P_i(s)$. Many existing $\mathcal{H}_2$/passive control design approaches (e.g., [42, 51]) cannot accommodate the performance specification given in (5.32), because they require the following assumptions:

$$B_1 D_{21}^T = 0 \text{ and } D_{12}^T C_1 = 0.$$ (5.48)

Here we present a method based on successive semidefinite programming (SSDP) techniques. To keep the control synthesis problem manageable, an *ad hoc* LQG control structure (an observer plus state feedback) similar to [42] is adopted:

$$k_i'(s) : \begin{cases} \dot{x}_c = Ax_c + B_2 u_i + L\left(y_i - C_2 x_c - D_{22} u_i\right) \\ u_i = -K_{gi}x_c, \end{cases}$$ (5.49)

This approach does not require the assumptions given in (5.48). As a result, $\mathcal{H}_2$ problems with any performance specification, such as that in (5.42), can be solved. From (5.49), controller $k_i'(s)$ should have the following state-space representation:

$$A_{ki} = A + B_2 K_{gi} - LC_2 - LD_{22}K_{gi}, \quad B_{ki} = L,$$
$$C_{ki} = K_{gi}, \qquad\qquad\qquad\qquad D_{ki} = 0. \tag{5.50}$$

where $L = \left(\Pi C_2^T + B_1 D_{21}^T\right)\left(D_{21}D_{21}^T\right)^{-1}$ is the observer gain matrix and Π is the solution to the Riccati equation below:

$$\Pi\left[A^T - C_2^T\left(D_{21}D_{21}^T\right)^{-1}D_{21}B_1^T\right] + \left[A - B_1 D_{21}^T\left(D_{21}D_{21}^T\right)^{-1}C_2\right]\Pi$$
$$- \Pi C_2^T\left(D_{21}D_{21}^T\right)^{-1}C_2\Pi + B_1 B_1^T - B_1 D_{21}^T\left(D_{21}D_{21}^T\right)^{-1}D_{21}B_1^T = 0. \tag{5.51}$$

The above $\mathcal{H}_2$ controller does not include integral action which is needed for offset-free control. This problem can be overcome by explicitly introducing integral action into the controller structure. For any passive controller $k_i'(s)$, $k_i'(s) + k_{si}/s$ is still passive when $k_{si} \geq 0$. Therefore, the inclusion of an integrator does not violate the stability condition. The structure for $k_i'(s)$ is now formed as follows:

$$A_{Iki} = \begin{bmatrix} A + B_2 K_{gi} - LC_2 - LD_{22}K_{gi} & 0 \\ 0 & 0 \end{bmatrix}, \quad B_{Iki} = \begin{bmatrix} L \\ 1 \end{bmatrix}, \tag{5.52}$$

$$C_{Iki} = \begin{bmatrix} K_{gi} & k_{si} \end{bmatrix}, \qquad\qquad\qquad D_{Iki} = 0.$$

The final controller for the ith loop,

$$k_i(s) = U_{ii}k_i'(s)\left[1 + w(s)k_i'(s)\right]^{-1}, \tag{5.53}$$

will retain the integral action as long as $w(0) = 0$, which implies that the process is decentralized integral controllable (DIC). Assume that

$$\mathcal{F}_l\left(P_i(s), k_i'(s)\right) := (A_{cl}, B_{cl}, C_{cl}, D_{cl}). \tag{5.54}$$

By using the positive-real lemma (Lemma 2.16) and property of system $\mathcal{H}_2$-norm, the above control problem can be cast into a matrix inequality problem:

Problem 5.12.

$$\min_{K_{gi}, k_{si}, P_1, P_2, Q} \{\mathrm{Tr}(Q)\}, \tag{5.55}$$

subject to

$$\begin{bmatrix} A_{ki}^T P_1 + P_1 A_{ki} & P_1 L - K_{gi}^T \\ L^T P_1 - K_{gi} & 0 \end{bmatrix} \leq 0, \tag{5.56}$$

$$k_{si} > 0, \tag{5.57}$$
$$P_1 > 0, \tag{5.58}$$

$$\begin{bmatrix} A_{cl}^T P_2 + P_2 A_{cl} & P_2 B_{cl} \\ B_{cl}^T P_2 & -I \end{bmatrix} < 0, \tag{5.59}$$

$$\begin{bmatrix} P_2 & C_{cl}^T \\ C_{cl} & Q \end{bmatrix} > 0, \tag{5.60}$$

where A_{ki} is defined in (5.50) and

$$A_{cl} = \begin{bmatrix} A & B_2 K_{gi} & B_2 k_{si} \\ LC_2 & A + B_2 K_{gi} - LC_2 & LD_{22}k_{si} \\ C_2 & D_{22}K_{gi} & D_{22}k_{si} \end{bmatrix}, \quad B_{cl} = \begin{bmatrix} B_1 \\ LD_{21} \\ D_{21} \end{bmatrix}, \quad (5.61)$$

$$C_{cl} = \begin{bmatrix} C_1 & D_{12}K_{gi} & D_{12}k_{si} \end{bmatrix}, \qquad\qquad D_{cl} = D_{11}.$$

This problem has four matrix decision variables P_1, P_2, Q, K_{gi} and one scalar variable k_{si}. Inequalities 5.56 to 5.58 imply passivity of $k_i'(s)$. Inequalities 5.59 to 5.60 represent the $\mathcal{H}_2$-norm condition. The trace of matrix Q gives the upper bound of $\|\mathcal{F}_l\left(P_i(s), k_i'(s)\right)\|_2^2$. Because A_{cl}, B_{cl}, C_{cl} and D_{cl} are functions of K_{gi} and k_{si}, (5.56) and (5.59) are bilinear matrix inequalities (BMIs). One way to solve the above problem numerically is to use the SSDP approach.

Approximation of bilinear constraints

One way to solve bilinear matrix inequalities is to approximate the bilinear terms using the Taylor expansion: Assume that X and Y are two independent matrix decision variables. Their product around (X_0, Y_0) is approximated by using the following equation:

$$XY = X_0 \delta Y + \delta X Y_0 + \delta X \delta Y + X_0 Y_0 \qquad (5.62)$$
$$\approx X_0 \delta Y + \delta X Y_0 + X_0 Y_0, \qquad (5.63)$$

where $\delta X = X - X_0$ and $\delta Y = Y - Y_0$. Both X and Y are restricted by their matrix norm:

$$\|\delta X\| \le \epsilon, \quad \|\delta Y\| \le \epsilon, \qquad (5.64)$$

where ϵ is an arbitrary small positive number such that the solution region of (5.63) is not too far from that of (5.62).

Define $\delta K_{gi} = K_{gi} - K_{gi0}$, $\delta k_{si} = k_{si} - k_{si0}$, $\delta P_1 = P_1 - P_{10}$ and $\delta P_2 = P_2 - P_{20}$. By using (5.63), (5.56) to (5.60) can be approximated around K_{gi0}, k_{si0}, P_{10} and P_{20}. This forms an approximated problem with the decision variables expressed as deviation values which is detailed in Section A.2 as Problem A.1. When solving the approximated problem, the solution radii of the deviation variables need to be restricted.

Initial solution

The iterative SSDP approach needs an initial solution such that all constraints in Problem 5.12 are satisfied. One obvious choice of the initial point is an arbitrary passive controller. With the assumption of $K_{gi} = L^T P_1$, the following inequality gives sufficient condition for (5.56):

$$X^T P_1 + P_1 X < 0, \tag{5.65}$$

where

$$X = A + B_2 L^T P_1 - LC_2 - LD_{22}L^T P_1. \tag{5.66}$$

With left and right multiplying of P_1^{-1}, and defining $W = P_1^{-1}$, the following LMI can be obtained:

$$W (A - LC_2)^T + (A - LC_2) W + LB_2^T + B_2 L^T - LD_{22}^T L^T - LD_{22}L^T < 0. \tag{5.67}$$

Matrix variable W can be solved using any semidefinite programming (SDP) tool, such as the MATLAB® *Robust Control Toolbox*. The state feedback gain matrix can be calculated as

$$K_{gi} = L^T W^{-1}. \tag{5.68}$$

The initial value of k_{si} can be set at an arbitrary positive value (*e.g.*, 1). With K_{gi} and k_{si} fixed and leaving P_1, P_2 and Q as decision variables, (5.56) to (5.60) become linear and thus can be solved using the SDP technique. The results can be used as the initial point for SSDP iteration.

The detailed SSDP iteration procedure for control design can be found in Section A.2 as Procedure A.2.

Control design procedure

Given a stable process $G(s) \in \mathbb{C}^{m \times m}$, a decentralized fault-tolerant $\mathcal{H}_2$ controllers can be synthesized by the following steps:

Procedure 5.13 (DUS $\mathcal{H}_2$ control design [16])

1. *Determine the pairing scheme for controlled and manipulated variables according to Procedure 5.2. If no pairings satisfy the steady-state passivity condition $\nu_D(G^+(s), 0) < 0$, the proposed decentralized detunable $\mathcal{H}_2$ controller design method will automatically lead to a DUS controller without integral action.*
2. *For the selected pairing scheme, obtain a stable and minimum phase scalar transfer function $w(s)$ such that $\nu_- (w(s), \omega) \leq -\nu_D (G^+(s), \omega)$, by solving Problem 5.10.*
3. *For each subsystem $G_{ii}^+(s)$ $(i = 1, \ldots, m)$, choose the weighting functions $w_{si}(s), w_{ki}$ and solve Problem 5.12 by implementing the proposed SSDP procedure to get the ith controller $k_i(s)$.*
4. *Obtain the multiloop controller $K(s)$ using (5.53).*

As an independent design approach, the performance design of DUS control is based on individual loops. Therefore, good $\mathcal{H}_2$ performance for the individual loops does not guarantee the performance of the overall plant, in particular if there are severe interactions between loops. As a result, simulation studies on closed-loop responses with the full process model need to be performed to check the overall performance. SDP techniques can deal with variables with structural constraints. Therefore, the above control synthesis approach can be extended to optimize the $\mathcal{H}_2$-norm of the overall system, with the decentralized control structure. In this case, individual controllers form a block diagonal control system, which has a block diagonal state feedback gain matrix:

$$K_g = \text{diag}\left\{K_{gi}\right\}, \quad i = 1, \ldots, m, \tag{5.69}$$

and a diagonal coefficient matrix for the integral term

$$K_s = \text{diag}\left\{k_{si}\right\}, \quad i = 1, \ldots, m. \tag{5.70}$$

In this case, the observers of controller loops can still be obtained from individual process loops. When matrix decision variables K_g and K_s are defined in Problem 5.12, the state feedback gain and integrator coefficients of all controller loops can be solved simultaneously to optimize for the $\mathcal{H}_2$-norm of the overall plant. However, this approach will be very computationally complex for processes with large dimensions. Consequently, Procedure 5.13 is often a more effective approach in practice.

Similarly, a DUS controller can also be designed such that $\mathcal{H}_\infty$ performance is achieved. In this case, the control problem is formulated as follows:

Problem 5.14. For all $i = 1, \ldots, m$

$$\min_{k_i'(s)} \left\{\gamma_i\right\}, \tag{5.71}$$

subject to

$$k_i'(s) \text{ is passive}, \tag{5.72}$$

and

$$\left\|T_i'\left(G_{ii}, k_i'\right)\right\|_\infty < 1. \tag{5.73}$$

The above problem can be solved by using the SPR/$\mathcal{H}_\infty$ control synthesis approach described in Section 3.5.

5.4.5 Illustrative Example

We consider the fault-tolerant control problem of a distillation column. This case study was first published in [16]. The process has the following transfer function [79]:

$$G(s) = \begin{bmatrix} \dfrac{2.22e^{-2.5s}}{(36s+1)(25s+1)} & \dfrac{-2.94(7.9s+1)e^{-0.05s}}{(23.7s+1)^2} & \dfrac{0.017e^{-0.2s}}{(31.6s+1)(7s+1)} & \dfrac{-0.64e^{-20s}}{(29s+1)^2} \\ \dfrac{-2.33e^{-5s}}{(35s+1)^2} & \dfrac{3.46e^{-1.01s}}{32s+1} & \dfrac{-0.51e^{-7.5s}}{(32s+1)^2} & \dfrac{1.68e^{-2s}}{(28s+1)^2} \\ \dfrac{-1.06e^{-22s}}{(17s+1)^2} & \dfrac{3.511e^{-13s}}{(12s+1)^2} & \dfrac{4.41e^{-1.01s}}{16.2s+1} & \dfrac{-5.38e^{-0.5s}}{17s+1} \\ \dfrac{-5.73e^{-2.5s}}{(8s+1)(50s+1)} & \dfrac{4.32(25s+1)e^{-0.01s}}{(50s+1)(5s+1)} & \dfrac{-1.25e^{-2.8s}}{(43.6s+1)(9s+1)} & \dfrac{4.78e^{-1.15s}}{(48s+1)(5s+1)} \end{bmatrix}. \tag{5.74}$$

Here we assume the pairing scheme of 1-1/2-2/3-3/4-4. The process with this pairing is DIC. Since the diagonal elements of $G(0)$ are all positive, $G^+(s) = G(s)$. The diagonally scaled passivity index is shown in Figure 5.11. The passivity index weighting function $w(s)$ was obtained by solving Problem 5.10:

$$w(s) = \frac{0.0397s\,(s + 14.078)}{(s + 0.760)\,(s + 0.0227)}. \tag{5.75}$$

The passivity index of $w(s)$ and $-\nu_D\,(G^+(s), \omega)$ are shown in Figure 5.10, from which it can be seen that $\nu_-\,(w(s), \omega) < -\nu_D\,(G^+(s), \omega)$.

The SSDP procedure was then applied to each loop to produce a multiloop $\mathcal{H}_2$ controller. The constraint given in (5.40) was imposed on all loops to guarantee DUS. While in theory different performance weighting functions can be used for different loops, identical weighting functions were used in this example for simplicity:

$$w_{ki} = 1 \times 10^{-4}, \tag{5.76}$$

$$w_{si}(s) = \frac{1}{s + 0.001}, \tag{5.77}$$

for $i = 1, \ldots, 4$. (The following parameters are required to perform the control design procedure detailed in Section A.2: $\epsilon_0 = 100$, $\zeta = 0.001$ and $\eta = 100$). The $\mathcal{H}_2$ performance of each loop together with the number of iterations are listed in Table 5.2. The transfer functions of the multiloop DUS controller are listed below:

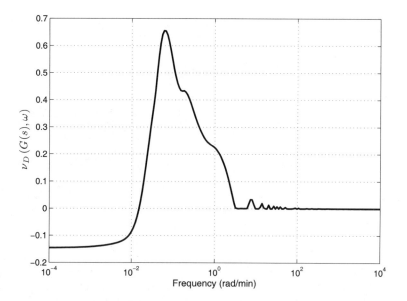

Fig. 5.10. Diagonally scaled passivity index

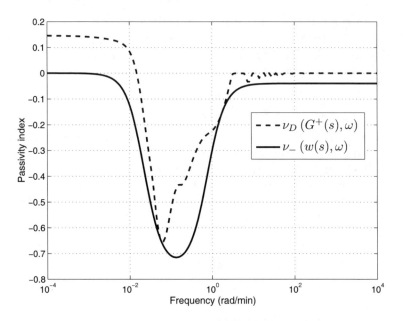

Fig. 5.11. Passivity indices of the process and weighting function $w(s)$ [16]

Table 5.2. $\mathcal{H}_2$ performance

Loop number	$\|\mathcal{F}_l(P_i, k_i')\|_2^2$	Number of iterations performed
1	19.385	16
2	3.563	18
3	1.977	21
4	6.178	7

$$k_1(s) = \frac{11.73s^5 + 55.77s^4 + 41.86s^3 + 5.088s^2 + 0.2552s + 0.003648}{s\left(s^5 + 5.425s^4 + 12.16s^3 + 26.72s^2 + 2.942s + 0.1185\right)}, \quad (5.78)$$

$$k_2(s) = \frac{62.42s^5 + 519.9s^4 + 1079s^3 + 644.5s^2 + 75.97s + 1.406}{s\left(s^5 + 10.92s^4 + 71.22s^3 + 302.1s^2 + 401.1s + 45.57\right)}, \quad (5.79)$$

$$k_3(s) = \frac{11.56s^5 + 41.72s^4 + 99.46s^3 + 121.2s^2 + 50.58s + 1.088}{s\left(s^5 + 4.085s^4 + 19.32s^3 + 37.35s^2 + 50.76s + 35.39\right)}, \quad (5.80)$$

$$k_4(s) = \frac{84.76s^4 + 593.1s^3 + 1094s^2 + 541s + 11.74}{s\left(s^4 + 10.48s^3 + 81.5s^2 + 328.1s + 380.3\right)}. \quad (5.81)$$

Simulation studies have been conducted with the proposed multiloop $\mathcal{H}_2$ controller. The integrals of the time-weighted absolute error (ITAE) of each loop without loop failures for the first 1000 minutes are as follows:

Loop 1: 1.369×10^4 Loop 2: 5.442×10^3
Loop 3: 2.606×10^4 Loop 4: 2.063×10^4

Figure 5.12 shows the transient responses of the controlled variables when step-change inputs are fed into all loops at time 0 and subsequently Loop 1 fails at the 400th minute. It demonstrates that the DUS controller maintains stability when loop failure occurs.

5.5 Fault-tolerant Control Design for Unstable Processes

Fault tolerance is often more important for control of unstable processes. Unstable processes are not decentralized unconditionally stabilizable and thus redundant controllers must be employed. The approach we present in this section is to stabilize the unstable process using a minimum number of single-loop static output feedback controllers and then implement the passivity-based DUS controller in the stabilized process. The static output feedback controller is built with redundancy and a fault detection and diagnostic (FDD) system. Static output feedback loops are used because they are simple and their backup units are cheaper to build and maintain. The performance design for the overall control system is performed during the (dynamic) DUS control

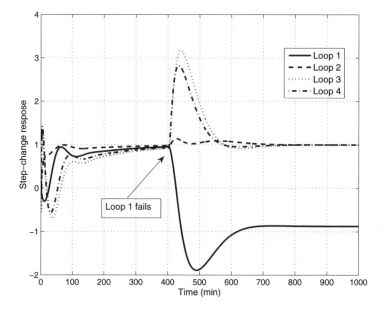

Fig. 5.12. Step-change responses [16]

synthesis (as described in the previous section). This section is based mainly on [17].

5.5.1 Static Output Feedback Stabilization

The key issue is to minimize the level of redundancy of the stabilizing controllers. For a given multivariable process, it is often unnecessary to implement a proportional controller for each input/output pair to stabilize the process. For example, the unstable multivariable continuous stirred-tank reactor (CSTR) with an exothermic reaction studied in [59] can be stabilized by a single proportional-only control loop that controls the reactor temperature. The single-loop stabilizing controllers collectively form a control system with a multiloop structure. The feasibility of the fault-tolerant control design hinges on the existence of a static stabilizing output feedback controller with structural constraints.

Static output feedback stabilizability

The necessary condition for static output feedback stabilizability was given by Wei [133]:

Definition 5.15 (Blocking zeros [156]). *Let $G(s) \in \mathbb{C}^{m \times m}$ be any proper real rational transfer matrix with the standard Smith-McMillan form:*

$$M(s) = \text{diag} \left\{ \frac{\alpha_i(s)}{\beta_i(s)} \right\}, \quad i = 1, \ldots, m. \tag{5.82}$$

A complex number $z_0 \in \mathbb{C}$ is called a blocking zero of $G(s)$ if $G(z_0) = 0$.

Definition 5.16 (Even parity-interlacing property (PIP) [133]). *An LTI multivariable system with a transfer function $G(s)$ in its Smith-McMillan form is said to satisfy the even PIP condition if*

1. *The number of real poles of $G(s)$ between any two real blocking zeros of $G(s)$ in the closed RHP is even.*
2. *The number of real blocking zeros of $G(s)$ between any two real poles of $G(s)$ in the closed RHP is even.*

Theorem 5.17 (Necessary condition for static output feedback stabilization [133]). *A necessary condition for static output feedback stabilizability of a given process $G(s)$ is that $G(s)$ satisfies the even PIP condition.*

Static output feedback control

The design problem of static output feedback controllers can be described as follows:

Problem 5.18 (Static output feedback stabilization with structural constraints). Given an LTI strictly proper process with the following transfer function and state-space representation:

$$G(s) : \begin{cases} \dot{x} = Ax + Bu \\ y = Cx, \end{cases} \tag{5.83}$$

where $x \in \mathbb{R}^n$, $u \in \mathbb{R}^m$ and $y \in \mathbb{R}^m$, find a static output feedback control law $u = Fy$, $F \in \mathcal{F}$, such that the closed-loop system

$$\Sigma_c : \begin{cases} \dot{x} = (A + BFC) x \\ y = Cx \end{cases} \tag{5.84}$$

is stable. $\mathcal{F} \subset \mathbb{R}^{m \times m}$ denotes the set of matrices with specified structures, *e.g.*, diagonal matrices or diagonal matrices with zero elements at pre-specified locations.

From the Lyapunov stability condition, the static output feedback stabilization problem is equivalent to the following feasibility problem:

Problem 5.19 ([17]). Find a matrix F with a specified structure and a Lyapunov matrix $P = P^T > 0$ such that the following matrix inequality is satisfied:

$$(A + BFC)^T P + P (A + BFC) < 0. \tag{5.85}$$

Because both P and F are decision variables, the above inequality is bilinear. Due to the structure constraints on matrix F, the variable transformation method developed in [42] cannot be used to convert (5.85) into an LMI. Here we adopt the approach based on the iterative linear matrix inequality (ILMI) method [27] with an extension so that the structural constraints on matrix variable F is dealt with properly. According to [27], (5.85) is satisfied if matrix variables

$$P > 0, \tag{5.86}$$

F and a scalar variable $\alpha < 0$ can be found such that the following LMI holds:

$$\begin{bmatrix} A^T P + PA - XBB^T P - PBB^T X + XBB^T X - \alpha P & (B^T P + FC)^T \\ B^T P + FC & -I \end{bmatrix} < 0, \tag{5.87}$$

where $X = X^T$ is any real positive definite matrix. Now Problem 5.18 is converted into a generalized eigenvalue problem. The static output feedback matrix F can be found by solving the following two problems alternately until $\alpha < 0$:

Problem 5.20 ([17]).

$$\min_{P, F} \{\alpha\},$$

subject to (5.86) and (5.87).

Problem 5.21 ([17]).

$$\min_{P,\,F} \left\{ \mathrm{Tr}(P) \right\},$$

subject to (5.86) and (5.87).

In both the above problems, the matrix decision variable F has a certain prespecified structure.

5.5.2 Fault-tolerant Control Synthesis

The entire fault-tolerant control system consists of two nested multiloop control subsystems, as shown in Figure 5.13. The inner loop is the static output feedback controller F built with redundancy, where F' is the backup controller. (Here we illustrate the configuration with one backup controller. It is sometimes necessary to implement more than one backup controller for mission critical processes). A fault detection and diagnostic system should be installed to monitor the main static feedback controller and switch the duty to the backup controller if the main controller fails. System $K(s)$ is a dynamic DUS controller which can be designed based on the stabilized process, following Procedure 5.13.

To reduce the level of redundancy, the static feedback controller F (and thus the backup controller F') should have the minimum number of nonzero diagonal elements. The minimum number of loops and corresponding structure of the static feedback controller can be determined by solving Problem 5.19

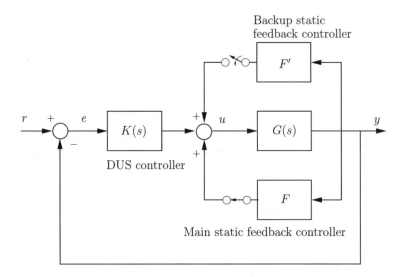

Fig. 5.13. Fault-tolerant control system with redundant static output feedback controller

with all possible (diagonal) structures of decision matrix variable F, using a different number of nonzero diagonal elements.

Given an unstable process modelled by (5.83), the decentralized fault-tolerant control system can be synthesized by the following procedure:

Procedure 5.22 (Fault-tolerant control design [17])

1. *Check the existence of a static output feedback stabilizing controller for a given unstable process by using the necessary condition in Theorem 5.17. If the condition is not satisfied, then this approach cannot be used.*

2. *Find the static multiloop output feedback controller $u = Fy$, that is, find a matrix F such that the condition in (5.85) is satisfied. This can be done by the following iterative procedure.*

   a) *Select $Q > 0$ and solve $P = P^T > 0$ from the following algebraic Riccati equation*

$$A^T P + PA - PBB^T P + Q = 0. \qquad (5.88)$$

   Set $X = P$.

   b) *Specify the structure of matrix F. It is preferable to start from a diagonal structure with the minimum number of nonzero diagonal elements.*

   c) *Solve Problem 5.20 for P, F and α. If $\alpha \le 0$, F is a stabilizing static output feedback gain. Go to Step 3.*

   d) *Solve Problem 5.21 for P and F using the α obtained in Step 2c.*

   e) *If $\|X - P\| > \delta$, a prescribed tolerance, set $X = P$ and go to Step 2c for the next iteration.*

   Otherwise, the static output feedback problem is not solvable by this ILMI approach with the user specified structure of F. Go to Step 2b and specify an alternative structure of the diagonal matrix F (by either swapping the positions of nonzero elements along the diagonal or replacing one diagonal zero with a nonzero parameter).

3. *Design the DUS controller $K(s)$ for the stabilized process $G'(s)$ following one of the approaches presented in Sections 5.4.1 and 5.4.2.*

Because the backup controllers are proportional-only (without any dynamics) and are identical to the main output feedback stabilizing controllers, bumpless transfer is automatically achieved whenever the duty is switched to the backup controllers in response to main control loop failures. The proposed control method can cope with multiple control failures that occur simultaneously, provided that the failures can be modelled by (5.1).

5.5.3 Illustrative Example

Consider the fault-tolerant control design problem of a CSTR [17]. An exothermic reaction $A \to B$ takes place and the heat is removed from the CSTR by external cooling water. The concentration of component A (c_A), component B (c_B) and the reactor temperature T are controlled by manipulating the feed flow rate F_f, the concentration of component A in the feed, c_{AF}, and the cooling water temperature T_c. The state-space representation of the linearised model is given as follows [59]:

$$\dot{x} = \begin{bmatrix} -0.1562 & 0 & -0.01553 \\ 0.0562 & -0.1 & 0.01553 \\ 0.7803 & 0 & 0.07958 \end{bmatrix} x + \begin{bmatrix} 0 & 0.1 & 0.1122 \\ 0 & 0 & -0.1124 \\ 0.0361 & 0 & -0.2 \end{bmatrix} u,$$

$$y = \begin{bmatrix} 1 & 0 & 0 \\ 0 & 1 & 0 \\ 0 & 0 & 1 \end{bmatrix} x,$$

(5.89)

where $x = y = \begin{bmatrix} C_A, C_B, T \end{bmatrix}^T$ and $u = \begin{bmatrix} T_c, C_{AF}, F_f \end{bmatrix}^T$. The system has an unstable pole of $s = 0.004$. Procedure 5.22 was implemented. It was easy to confirm that the process satisfies the even-PIP condition given in Theorem 5.17, which indicates that it is possible to stabilize the process using static output feedback. By solving Problems 5.20 and 5.21 iteratively, it was found that this process can be stabilized by either of the following two single-loop output feedback controllers: $F_1 = \text{diag}\{0, -0.1497, 0\}$ or $F_2 = \text{diag}\{0, 0, -0.0841\}$, which represent the least amount of redundant control loops required. In this case study, the second scheme was finally chosen because it requires a smaller controller gain.

The next step was to find a DUS controller for the stabilized process. The pairing study of the stabilized process showed that the diagonal pairing 1-1/2-2/3-3 is the only suitable choice because the other two pairing schemes lead to non-DIC process models. Based on this pairing, a decentralized unconditional stabilizing PI controller was designed following Procedure 5.7. The controllers have the following transfer functions:

$$k_1(s) = -0.327 \left(1 + \frac{1}{18s} \right),$$

$$k_2(s) = 0.2 \left(1 + \frac{1}{60s} \right) \quad \text{and}$$

(5.90)

$$k_3(s) = 0.01 \left(1 + \frac{1}{50s} \right).$$

To test system stability against control loop failures, simulation studies were carried out with a series of failure events. Controller F was assumed to fail at the 3000th minute and the control duty was switched to the backup controller F' at $t = 3005$ minute. The step-change responses of controlled variables to the stabilizing controller failure are shown in Figure 5.14. Figure 5.15 shows the step-change responses of controlled variables when the following control failures occurred: (1) The control valve for the cooling water (Loop 1) was stuck at $t = 1500$ minute. (2) The second loop of the PI DUS

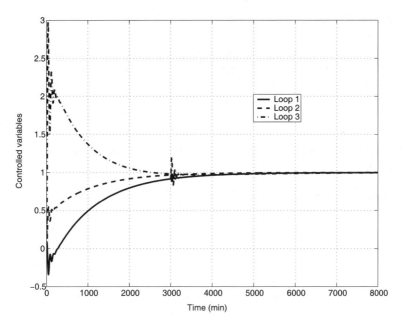

Fig. 5.14. Responses to failure of stabilizing controller [17]

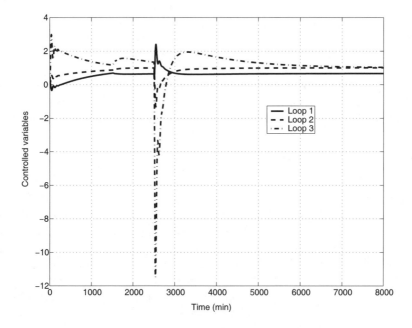

Fig. 5.15. Responses to valve stickiness and partial control failure [17]

controller, $k_2(s)$, partially (50%) failed at $t = 2500$ minute. (Note that the controlled variables in both figures are deviations from nominal values.) It can be seen that the closed-loop system remains stable when multiple control failures occur simultaneously. If integral action is used in each loop, when one control loop fails, the controlled variables of other loops are affected but will automatically return to their set points with the proposed control.

5.6 Hybrid Active-Passive Fault-tolerant Control Approach

The DUS control approach introduced in previous sections is a simple framework that guarantees system stability under control loop failure. However, as a redundant-free passive fault-tolerant strategy, DUS controllers alone may not be able to provide satisfactory performance when faults occur. As discussed in Section 5.1, the main issues of the active approaches are (1) they could be extremely complex and expensive to design, build and maintain, especially for the approaches with redundancy; (2) their reliability solely depends on fault-free and efficient running of the fault detection systems. In this section, we introduce a hybrid approach that combines the DUS framework with active fault-tolerant control approaches (with or without redundancy), largely based on and with an extension to [122]. In this approach, only the failure events that often occur or have significant impact on process operation are detected and accommodated by an active FTC while all other possible failure events are handled by the DUS controller. This approach can significantly reduce the complexity and improve the cost effectiveness of the FTC system.

This hybrid active-passive FTC system may consist of a fault detection and diagnosis subsystem, backup control loops (with backup sensors and actuators), a set of controllers predesigned for specific faults and a DUS controller.

5.6.1 Failure Mode and Effects Analysis

In this FTC approach, it is important to identify the "significant" fault events, which either have high probability of occurrence and/or may lead to severe consequences for process operation. A commonly accepted approach that can be adopted for this purpose is the failure modes and effects analysis (FMEA), which was originally developed by reliability engineers. It includes finding, for each main component (*e.g.*, sensors and actuators), the failure mode (*i.e.*, in what way it can fail), failure cause, failure effect and failure probability. More details can be found in [20]. This analysis will help determine the failure events that should be detected and isolated by the FDD subsystem and properly handled by an active fault-tolerant controller (with or without backup loops).

5.6.2 Fault Detection and Accommodation

There are quite a few existing fault detection and diagnosis approaches. A comprehensive treatment can be found in [20]. Most of these approaches can be used in the hybrid active-passive FTC framework. Here we illustrate the hybrid active-passive FTC approach using the residual method similar to [83, 142]. It uses the process model to predict the process output $\hat{y}(t)$ for the measured input $u(t)$. The difference between the actual process output $y(t)$ and $\hat{y}(t)$ is called the residual:

$$\mu(t) = y(t) - \hat{y}(t). \tag{5.91}$$

In the faultless case, the residual vanishes or is close to zero. It becomes significantly large in response to faults, disturbances, modelling errors and estimation errors of initial conditions. The FDD system tests the residuals against thresholds that are selected theoretically or empirically to determine possible faults. For sensor fault detection, the well-known Kalman filter can be employed in residual generation to reduce the effects of noise and the estimation errors of initial state values. Consider an LTI process:

$$\begin{aligned} \dot{x} &= Ax + Bu + Gw, \\ y &= Cx + Du + v, \end{aligned} \tag{5.92}$$

with control inputs u, process noise w and measurement noise v. The noise covariance matrices are

$$E\{ww^T\} = Q_m, \quad E\{vv^T\} = R_m \quad \text{and} \quad E\{wv^T\} = N_m. \tag{5.93}$$

When a sensor fault happens, the output equation of (5.92) becomes

$$y_f = C_f x + D_f u + v_f, \tag{5.94}$$

where y_f is the output vector y with the element(s) corresponding to the faulty sensor(s) set to zero. Matrices C_f and D_f can be obtained by setting the rows in matrices C and D that correspond to the faulty sensor(s) to zero vectors. There is one pair of matrices C_f and D_f for each "significant" sensor fault scenario, including the combination of different sensor faults. A bank of separate Kalman filters is designed based on the faultless (normal) case and a number of possible failure situations. The nominal filter for normal operation is as follows [122]:

$$\dot{\hat{x}} = A\hat{x} + Bu + K_0(y - C\hat{x} - Du). \tag{5.95}$$

The filter for the ith fault scenario is given by

$$\dot{\hat{x}}_{f,i} = A\hat{x}_{f,i} + Bu + K_{f,i}(y_{f,i} - C_{f,i}\hat{x}_{f,i} - D_{f,i}u), \tag{5.96}$$

where $\hat{x}$ and $\hat{x}_{f,i}$ are the estimate of states. Matrices K_0 and $K_{f,i}$ are the Kalman estimator gains, which can be calculated by solving the algebraic

Riccati equation for Kalman filters. If there are n_f failure scenarios which need to be isolated, a bank of $(n_f + 1)$ Kalman filters should be implemented simultaneously. The error of output estimation can be calculated as the residuals:

- Filter for faultless operation:

$$\nu_0 = y - C\hat{x} - Du. \tag{5.97}$$

- Filter i ($i = 1, \dots, n_f$):

$$\nu_i = y_{f,i} - C_{f,i}\hat{x}_{f,i} - D_{f,i}u. \tag{5.98}$$

If no sensor fault occurs, all Kalman filters should be zero-mean, Gaussian innovation processes with limited covariance. When the ith fault scenario occurs (assuming the process is still observable in this scenario), all but the ith Kalman filter will produce a residual that is no longer vanishing or near zero, and the covariance will increase because of the erroneous measurement data. The sensor failure scenarios can be isolated by using either a hypothesis test if the occurrence probability of each failure scenario is known (*e.g.*, [83]), or a deterministic filter. In the latter case, an exponentially weighted moving average filter within a time interval ΔT can be used (as in [5]):

$$\mu_i = \int_{(k-1)\Delta T}^{k\Delta T} e^{-\lambda(t-\tau)} \|\nu_i\|^2 \, d\tau, \quad \forall \, i = 0, \dots, n_f, \tag{5.99}$$

where λ is a positive smoothing constant. The sensor failure scenarios (one scenario can have several faulty sensors) can be detected and isolated by comparing directly μ_i with a predetermined threshold $\bar{\mu}_i$:

- If $\mu_i < \bar{\mu}_i \; \forall \, i = 0, \dots, n_f$, then no fault occurs.
- If $\mu_i < \bar{\mu}_i$ and $\mu_j > \bar{\mu}_j \; \forall \, j = 0, \dots, n_f, \; j \neq i$, then fault scenario i is detected.

In any other situations, *e.g.*, where more than one residual is smaller than its threshold, an unisolated fault scenario may have occurred. In this case, the active FTC is inactive and an alarm is set off, requesting engineers to investigate the problem. The control system stability is maintained by the DUS controller. The above FDD approach is effective when the states of the process are observable from the remaining healthy sensors, that is, $\{C, A\}$ and $\{C_{f,i}, A\}$ ($i = 1, \dots, n_f$) are observable. In some cases, explicitly or implicitly redundant sensors should be installed such that the above observability condition is satisfied. In the latter case, some sensors measure different process variables which are dependent on each other. Most industrial processes are equipped with implicit redundant sensors so that the operators can cross-check the feedback from sensors.

The above FDD method can also be applied to detect actuator faults, where the state equation of (5.92) becomes

$$\dot{x} = Ax + B_f u + Gw, \tag{5.100}$$

where the column of matrix B_f corresponding to the faulty actuator is set to a zero vector. Actuator faults can also be detected by the actuator built-in feedback (e.g., the valve position feedback from a control valve) or sensors that measure actuator outputs directly or indirectly.

Once a fault scenario is detected by the above FDD system, the estimated state information from the Kalman filter can be used to reconstruct the virtual sensor output. If the ith fault scenario is detected, then the virtual sensor output can be calculated based on the states in the ith FDD filter, i.e.,

$$\begin{aligned} \dot{\hat{x}}_{f,i} &= A\hat{x}_{f,i} + Bu + K_{f,i}(y - C_{f,i}\hat{x}_{f,i} - D_{f,i}u), \\ \hat{y} &= C\hat{x}_{f,i} + Du. \end{aligned} \tag{5.101}$$

5.6.3 Control Framework

Figure 5.16 shows an example of a hybrid FTC system consisting of a DUS controller, a FDD unit, backup actuators and virtual sensors. The FDD unit detects possible actuator and sensor faults. If an actuator fault is detected, a backup actuator will be implemented. If a sensor fault is detected, the virtual sensor output is used instead of the actual faulty sensor. The control design procedure is as follows:

Procedure 5.23 (Hybrid FTC design)

1. *Conduct the failure mode and effect analysis to identify the sensor/actuator faults which have high occurring rates or have significant impact on process operation. This requires prior knowledge of sensor/actuator failure rates and past process operating data, in particular, statistics of past failure events.*
2. *When the state observability of the process is not satisfied under some sensor failures which need to be detected and isolated, implicitly or explicitly redundant sensors need to be added to maintain state observability in these fault scenarios. If a faulty actuator may have significant impact on process operation or cause system instability, a backup actuator should be implemented.*
3. *Design FDD filters for the fault scenarios identified in Step 1 according to Section 5.6.2.*
4. *Design the virtual sensors using (5.101).*
5. *Design a DUS controller to deal with failures that are not detected or not significant, following Procedure 5.13.*

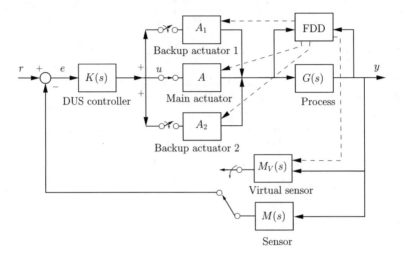

Fig. 5.16. Hybrid FTC system

5.7 Summary

In this chapter, we have discussed the framework of passivity-based fault-tolerant control. It is built on the decentralized unconditional stability (DUS) condition. We can use this framework to design fault-tolerant controllers that are redundant-free for stable processes or with a low level of redundancy for unstable processes. In practice, the DUS control should be combined with fault detection and diagnosis and fault accommodation techniques to deliver cost-effective fault-tolerant control.

6

Process Controllability Analysis Based on Passivity

Process control has been playing an important role in process industries as increased process integration and tight operating conditions are putting greater demands on control system performance. For a given process system, the control performance achievable can be quantified by the input-output controllability measure. Controllability analysis can be used in the process design stage to reveal controllability problems. In this chapter, we will introduce a controllability analysis approach based on passivity.

6.1 Introduction

The traditional approach to process design and control has been to design the process and the control systems sequentially. In the first stage, the design engineer constructs a process flowsheet that optimizes the economics of the project, including steady-state operating and capital costs. The control engineer must then devise the control systems to ensure stable dynamic performance and to satisfy the operational requirements [78]. Because little consideration is given to dynamic controllability in this procedure, the outcome of this approach sometimes is a plant whose dynamic characteristics lead to severe operating problems and significant economic penalties [80, 93].

For a given process design, dynamic controllability determines whether the process can be controlled effectively by a feedback control system. Such controllers need to reject disturbances at a specified steady state for reduced product variability, and/or to move the process fast and smoothly from one operating condition to another, including start-up and shutdown [94, 134]. The definition of input-output controllability in this chapter is similar to process resilience [86] and switchability [137], rather than the often referred to state controllability of Kalman.

Consideration of dynamics and control during process design could have significant economic impact [34]. It is well known that a process design fundamentally determines its inherent controllability, because it imposes inherent limitations on control performance regardless of the control method implemented. Therefore process controllability analysis should be performed in the process design stage. This is particularly important when plants become more complex with a large number of controlled and manipulated variables, and/or integrated with recycle streams and energy integration.

In the past two decades, integration of process design and control has drawn considerable interest in both industry and academia. One approach is to solve the design/control integration problems using optimization techniques (e.g., [71, 94]). To deal with continuous decision variables, such as process parameters, and discrete (integer) decision variables, such as alternative process and control structures, the design problems are mathematically represented as large-scale, mixed-integer nonlinear programming (MINLP) problems (e.g., [7, 108]). The optimization-based approach makes it possible to explore economic trade-offs and to handle nonlinear models directly. Ideally, this approach could solve the problems of process design and control design simultaneously. However, the amount of computation that the optimization approach requires may be extremely high. To make the problem manageable, some existing methods assume that the controlled variables are perfectly controlled at set points and perform optimization without operating variables [94]. Some other methods assume a linear time invariant control structure [71]. These approaches may lead to an unrealistic estimate of the achievable performance. Even with these simplifications, the optimization methods are still very computationally complex and thus their applications are currently restricted to small scale problems [93].

Another approach is to undertake the controllability analysis based on the open-loop characteristics of processes. Manipulated variable constraints, nonminimum phase behavior, nonlinearity, loop interactions and model–plant mismatch all impose limitations on process controllability [56, 86]. These effects have been analysed and assessed by using certain open-loop indicators, such as minimum singular values, RHP zeros, time delays and condition numbers (e.g., [18, 86]). However, the above analysis methods suffer from the following weaknesses: (1) They are based on linear models and thus are suitable only for linear or mildly nonlinear processes; and (2) they suggest only the likely effect of each attribute on the closed-loop performance but fail to indicate the overall effect of the characteristics on dynamic controllability. To deal with nonlinear models and control constraints, the operability index (OI) method [132] was developed based on operating spaces, which, for a given static nonlinear model, addresses controllability issues associated with multiplicity and mapping from the input space to the output space. This method was extended to dynamic operability analysis by studying the dynamic operating spaces achieved within the desired response time [129]. However, these extensions require solving very computationally intensive nested or iterative optimization problems.

Passive systems (both linear and nonlinear) represent a class of minimum phase systems, which are very easy to control, even if they are highly nonlinear and/or highly coupled. As shown in Chapter 2, a passive system can be stabilized by a simple static output feedback control law with an arbitrarily high gain, provided that the zero-state detectability condition is satisfied. We have also seen in Chapters 2 and 3 that the frequency-dependent passivity index reflects the total destabilizing effect of RHP zeros, time delays and coupling and may indicate the dynamic performance achievable for linear systems. The integral controllability results we have seen in Chapter 4 are examples of passivity-based controllability analysis for decentralized control systems. Most importantly, the concept of passive systems and the Passivity Theorem are also valid for nonlinear systems. Intuitively, the passivity index can be used to infer its controllability. As such, passivity-based controllability analysis can be applied to nonlinear processes, where the characteristics of linear systems such as phase angles and RHP zeros are no longer applicable. In the next section, we will develop a framework for passivity-based controllability analysis.

6.2 Analysis Based on Extended Internal Model Control

In this section, we extend the internal model control (IMC) framework so that the concept of passive systems can be used in controllability analysis. We also illustrate how the extended framework is used in controllability analysis for linear and nonlinear processes.

6.2.1 Extended Internal Model Control Framework

Consider a feedback control system with a stable process (G) and a controller (K), as shown in Figure 6.1a (assuming $G_{ff} = 0$). Here $G : u \mapsto y$ and $K : e \mapsto u$ are multivariable nonlinear operators. The mappings are denoted as $y = Gu$ and $u = Ke$, where all the variables are vectors. The dynamic performance, in terms of disturbance attenuation and set point tracking, can be represented by the dynamic systems from the disturbance (d) to control error (e) and from reference (r) to controlled output (y), respectively. If the process G is stable and strictly input passive (here we adopt the input-output version given in Definition 2.22), according to Theorem 2.46, the closed-loop system will be $\mathcal{L}_2$ stable if any passive controller K is employed. Such controllers could have arbitrarily large gains to achieve "perfect control" if there are no constraints on the manipulated variables. If G lacks IFP, then the controller must have excessive OFP to ensure closed-loop stability. This implies a performance limit because a system with excessive OFP has limited $\mathcal{L}_2$ gain. Assume that a dynamic feedforward system G_{ff} can be found such that $G_p = G + G_{ff}$ is strictly input passive, then the passivated system G_p can be stabilized by any passive controller. By loop shifting, it can be seen that the negative feedback system of K and G_{ff} forms a stabilizing controller C for the original system G, as shown in Figure 6.1b.

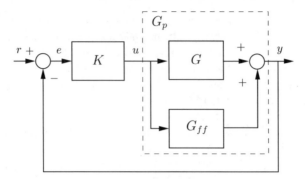

(a) Input feedforward passivation

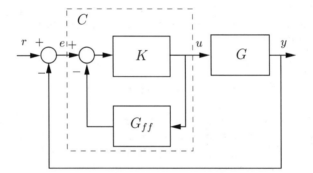

(b) Final controller C

Fig. 6.1. IFP and controllability

The above feedback system can be presented in an extended IMC structure [120]. The IMC structure is shown in Figure 6.2, where G is the process system, $\tilde{G}$ is the process model and Q is the so-called IMC controller. Assuming no model–plant mismatch, *i.e.*, $\tilde{G} = G$, and the initial conditions of systems $\tilde{G}$ and G are identical, the closed-loop relation between the process output $y(t)$ and the reference signal $r(t)$ is affine with respect to Q:

$$y(t) = GQr(t), \qquad (6.1)$$

where GQ represents systems Q and G connected in series. Therefore, if G is stable, the condition for internal stability of the closed-loop control system is determined by the stability of the IMC controller Q. Clearly if G is invertible, then perfect control is possible by using an IMC controller $Q = G^{-1}$. If G is not invertible but an invertible approximation can be found, then the best performance can be achieved by an IMC controller that inverts the invertible approximation of G. This shows that

1. Any feedback controller provides an approximate inverse of the plant model.

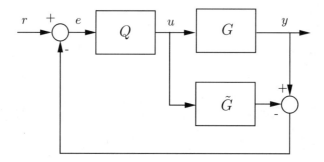

Fig. 6.2. Internal model control framework

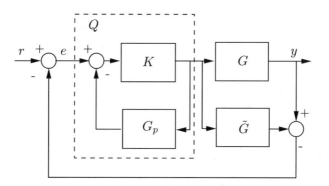

Fig. 6.3. Extended IMC framework

2. Control performance is limited by the invertibility of the process system.

For chemical process control, typical noninvertible components are non-minimum phase (NMP) elements that generally include time delays and RHP zeros. It is well known that process NMP characteristics may impose severe performance limitations on closed-loop systems [114]. Therefore, IMC provides an open-loop framework for closed-loop system controllability analysis [86]. Several controllability analysis approaches have been developed for linear process systems based on IMC [86, 134, 153]. For example, in [153], a stable linear process model $G(s)$ is factorized into a minimum phase part $M(s)$ and a nonminimum phase part $N(s)$ using all-pass factorization. The minimum phase part is invertible and it was proven that the IMC controller $Q(s) = M^{-1}(s)$ achieves optimal quadratic control performance if the controller gain is not limited. Therefore, the best performance achievable can be quantified by the smallest sensitivity function $S(s) = I - N(s)$. However, such factorization is very difficult (if not impossible) to extend to nonlinear systems.

The IMC framework can be extended using the concept of passivity. Denote $\tilde{G}$ as the process model and G_p as the passivated system of $\tilde{G}$ (e.g., using

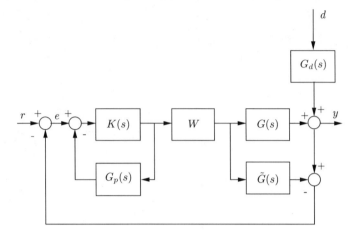

Fig. 6.4. Controllability analysis for linear processes

the passivation methods given in Section 2.5). The passivity-based control shown in Figure 6.1 can be represented using the IMC structure depicted in Figure 6.3, where the IMC controller Q is the feedback system of a passive system K and G_p. Because G is stable, the closed-loop stability is determined by the stability of the IMC controller Q. Since G_p is strictly input passive and stable, the stable IMC controller Q can be parameterized by any passive K, according to the Passivity Theorem. When a passive system K is chosen to have infinite gain, Q approaches the inverse of G_p. Therefore, the best performing closed-loop system from r to y, subject to closed-loop stability, can be estimated by GG_p^{-1}, independent of the choice of controller. The system from reference r to control error e is $I - GG_p^{-1}$. The passivated process G_p can be regarded as the passive approximation of G. In this case, process G is decomposed into a passive subsystem G_p and a nonpassive subsystem $(-G_{ff})$ (as in [121]). Because the passive approximation of a process (G_p) is minimum phase and always invertible (additional dynamics may be required to make the inverse causal), the IFP index of the process indicates its invertibility. Unlike the factorization approach in IMC-based controllability analysis, passivity-based decomposition is applicable to both linear and nonlinear processes. Note that the final controller C includes K and G_{ff}. Therefore, the above controller parameterization is not limited to passive final controllers. From the above discussion, it can be seen that *input-output controllability can be inferred from the input feedforward passivity of the process.*

6.2.2 Controllability Analysis for Stable Linear Processes

Here we study controllability in terms of both disturbance rejection and set point tracking performance. It is assumed that

$$y(s) = G(s)u(s) + G_d(s)d(s), \qquad (6.2)$$

where the process system $G(s) \in \mathbb{C}^{m \times m}$, controller output $u(s) \in \mathbb{C}^m$, process output $y(s) \in \mathbb{C}^m$, disturbance $d(s) \in \mathbb{C}^p$ and the disturbance transfer function $G_d(s) \in \mathbb{C}^{m \times p}$. $K(s)$ is an arbitrary passive controller. The feedback system is shown in Figure 6.4. A constant matrix $W \in \mathbb{R}^{m \times m}$ is used to rescale the transfer function $G(s)$ such that

$$G(0)W + W^T G^T(0) > 0. \tag{6.3}$$

One possible choice is $W = G^{-1}(0)$ if $G(0)$ is nonsingular. $G_p(s) = G(s)W + G_{ff}(s)$ is the passive approximation of the scaled process. Matrix W will be absorbed into the final controller. Therefore, when $K(s) \to \infty$, the closed-loop system is $y(s) = G(s)WG_p^{-1}(s)r(t)$, leading to a sensitivity function (the system from $r(t)$ to $e(t)$):

$$S(s) = I - G(s)WG_p^{-1}(s). \tag{6.4}$$

Similarly, the disturbance rejection achievable can be represented by the transfer function from $d(s)$ to $y(s)$:

$$S_d(s) = S(s)G_d(s). \tag{6.5}$$

The dynamic tracking and regulating performance achievable can be quantified by $\bar{\sigma}(S(j\omega))$ and $\bar{\sigma}(S_d(j\omega))$, respectively. The largest frequency ω where $\bar{\sigma}(S(j\omega))$ first crosses -3dB (≈ 0.707) from below is called the closed-loop bandwidth, denoted as ω_B [114]. Up to ω_B, $\bar{\sigma}(S(j\omega)) < 0.7$ and the controller is effective in improving performance. At frequencies higher than ω_B, the controller is either ineffective (when $\bar{\sigma}(S(j\omega)) \approx 1$) or actually degrades performance (when $\bar{\sigma}(S(j\omega)) > 1$). A similar comment can be made for $S_d(s)$, where the combined effects of the feedback controller and the disturbance transfer function $G_d(s)$ are evaluated. A good choice of manipulated/controlled variables often leads to small $G_d(s)$, which implies small impact of the disturbance variables. The larger the bandwidth of the sensitivity function, the faster the dynamic response that the feedback control system can achieve.

The key issue here is to find the passive approximation. A frequency domain controllability approach using the above extended IMC framework was developed in [124, 125]. Here we present a state-space approach, which is more intuitive. Assume that the scaled system is given by $G(s)W := (A_g, B_g, C_g, D_g)$ and the passivated system is given by $G_p(s) := (A_p, B_p, C_p, D_p)$. Because feedforward passivation needs only to change the zero dynamics of $G(s)W$,

$$A_p = A_g, \quad B_p = B_g. \tag{6.6}$$

According to the positive-real lemma (Lemma 2.16), matrices C_p and D_p can be found such that (2.42) is satisfied to ensure the passivity of $G_p(s)$. To achieve offset-free control, we need $G_p(0) = G(0)W$ (i.e., $G_{ff}(0) = 0$). Therefore the following condition should be satisfied:

$$D_p = (C_p - C_g) A_g^{-1} B_g + D_g. \tag{6.7}$$

Such a D_p can always be found if $G(0)$ is nonsingular. The passive approximation can be found by solving the following LMI problem:

Problem 6.1.

$$\min_{P,C_p} \gamma,$$

subject to

$$P > 0, \tag{6.8}$$

$$\begin{bmatrix} A_g^T P + P A_g & P B_g - C_p^T \\ B_g^T P - C_p & D \end{bmatrix} < 0, \tag{6.9}$$

$$\begin{bmatrix} \gamma^2 I & C_p - C_g \\ C_p^T - C_g^T & I \end{bmatrix} > 0, \tag{6.10}$$

where $D = - \left[(C_p - C_g) A_g^{-1} B_g + D_g \right]^T - \left[(C_p - C_g) A_g^{-1} B_g + D_g \right]$.

Inequalities 6.8 and 6.9 come directly from the positive-real lemma. Inequality 6.10 represents the constraint on the matrix norm:

$$\|C - C_p\| < \gamma. \tag{6.11}$$

By minimizing the above matrix norm, a passive approximation of $G(s)W$ can be found. Alternatively, one may choose to minimize $\|G_{ff}(s)\|_\infty$. In this case, the following problem should be solved:

Problem 6.2.

$$\min_{P,C_p} \gamma,$$

subject to (6.8) and (6.9) and

$$X > 0, \tag{6.12}$$

$$\begin{bmatrix} A_g^T X + X A_g & X B_g & C_p^T - C_g^T \\ B_g^T X & -\gamma I & \left[(C_p - C_g) A_g^{-1} B_g \right]^T \\ C_p - C_g & \left[(C_p - C_g) A_g^{-1} B_g \right] & -\gamma I \end{bmatrix} < 0. \tag{6.13}$$

Inequalities 6.12 and 6.13 represent the following infinity-norm constraint:

$$\|G_{ff}(s)\|_\infty < \gamma. \tag{6.14}$$

To illustrate passivity-based controllability analysis, we consider the control scheme selection problem for a binary distillation column as follows:

Table 6.1. Dynamic controllability of binary distillation column

Configuration	ω_B (rad/min)	Settling time (min)
$D - V$	0.028	164.3
$RR - V$	0.019	242.1
$R - V$	9.1×10^{-4}	5054.9

Example 6.3 (Binary distillation column control). Consider the binary distillation column control problem studied in [146]. The process has a single feed stream and bottom and overhead streams as product streams. Candidates for manipulated variables are reflux ratio (RR), reflux flow (R), distillate flow (D) and vapor boilup (V). The controlled variables are the distillate composition (x_D) and bottom composition (x_B). The purpose of this study is to determine the controllability of different control schemes (with different manipulated variables). The process models for different configuration are given as follows [146]:

$$G_{DV}(s) = \begin{bmatrix} \dfrac{-2.3}{(29.4s+1)(0.2s+1)} & \dfrac{1.04e^{-19s}}{(1.35s+1)} \\ \dfrac{3.1e^{-1.5s}}{(19s+1)^2} & \dfrac{1.8(77s+1)}{(21.2s+1)(0.9s+1)} \end{bmatrix}, \qquad (6.15)$$

$$G_{RV}(s) = \begin{bmatrix} \dfrac{16.3}{(18.1s+1)} & \dfrac{-18}{(21.4s+1)} \\ \dfrac{-26.2e^{-1.4s}}{(47.6s+1)(1.2s+1)} & \dfrac{28.63}{(47.6s+1)(4.4s+1)} \end{bmatrix}, \qquad (6.16)$$

$$G_{RRV}(s) = \begin{bmatrix} \dfrac{1.5}{(34.2s+1)(4.1s+1)} & \dfrac{-1.12}{(1.4s+1)} \\ \dfrac{-2.06e^{-2.3s}}{(23.6s+1)^2} & \dfrac{4.73}{(19.7s+1)(0.7s+1)} \end{bmatrix}. \qquad (6.17)$$

By solving Problem 6.2, the maximum singular values of the achievable sensitivity functions for different configurations are obtained, as shown in Figure 6.5. The closed-loop bandwidth and approximate settling time of all configurations are summarized in Table 6.1, from which it can be seen that the $(D - V)$ configuration has the largest bandwidth and configuration $(R - V)$ gives the worst performance achievable.

The qualitative ranking of dynamic controllability is confirmed by the closed-loop dynamic simulation with multiloop PI controller tuned using the biggest log-modulus tuning (BLT) method [146].

As mentioned earlier, $W = G(0)^{-1}$ is a good choice for the scaling matrix in most cases. An optimal rescaling matrix W can also be found such that the frequency range in which $G(j\omega)W$ is positive real is maximized by solving the following problem:

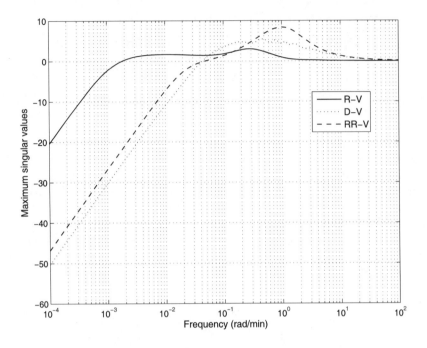

Fig. 6.5. Achievable sensitivity function

Problem 6.4 ([105]).

$$\max_W \omega,$$

subject to

$$G(0)W + W^T G(0)^T > 0, \tag{6.18}$$

$$G(j\omega)W + W^T G(j\omega)^\star > 0. \tag{6.19}$$

For nonlinear process systems, a passive approximation can be obtained by using the input feedforward passivation approach presented in Section 2.5.1, if the storage function of the nonlinear process is known. In this case, a nonlinear system from r to e and from d to y can be obtained. Simulation studies can be performed to analyse process dynamic controllability. This approach, however, requires an analytical (rather than numerical) approach and thus can be very complex and difficult to implement for general nonlinear process systems (*e.g.*, finding storage functions for arbitrary nonlinear systems can be a formidable task itself!). In the next few sections, we limit the scope to the steady state or simple dynamics so that a numerical approach can be developed for controllability analysis for nonlinear processes.

6.3 Regions of Steady-state Attainability

Controllability analysis for full nonlinear general dynamic models is often very difficult, mainly for two reasons: (1) such an analysis is very complex; (2) information on the process dynamics is usually very limited. Steady-state controllability analysis tools are very useful because information from steady-state simulations, material balance calculations as well as plant operation (plant historical data, plant step tests, *etc.*) can be used for analysis [132]. Many existing steady-state analysis approaches are limited to linear/linearised process models. For example, the RGA (the ratio of the open-loop gain for a specific pairing of manipulated variables when all other loops are closed [23]) shows the degree of interaction between controlled variables. Analysis based on singular value decomposition (*e.g.*, the singular value analysis [114] and condition number [18]) reveals the magnitude of the impact that manipulated and disturbance variables have on selected controlled variables.

The use of steady-state nonlinear models to assess the static operability of a process has recently been suggested by Vinson and Georgakis [132]. The inputs of the process can vary over a certain range, which is called the available input space (AIS). The output space that can be reached using the entire AIS, as calculated from the steady-state process model, is referred to as the achievable output space (AOS) (the AOS can be a function of both the manipulated variables and disturbance). The desired input space (DIS), which is a set containing the required values of the manipulated variables can be determined from desired values of the controlled variables, which form a desired output space (DOS). A set of *operability indices* (OI) are defined as the hypervolume of the intersection between the AOS, DOS, AIS and DIS:

$$\mathrm{OI}_y \triangleq \frac{\mu[\mathrm{AOS} \cap \mathrm{DOS}]}{\mu[\mathrm{DOS}]}, \tag{6.20}$$

$$\mathrm{OI}_u \triangleq \frac{\mu[\mathrm{AIS} \cap \mathrm{DIS}]}{\mu[\mathrm{DIS}]}, \tag{6.21}$$

where $\mu[\cdot]$ is the measure of the hypervolume of the corresponding regions (in a one-dimensional case, $\mu[\cdot]$ measures length; in a two-dimensional case area; in a three-dimensional case volume, *etc.*). For example, a process with an OI_y of 1 is a process where every steady-state output in the DOS can be reached using an input action within the AIS. The above OI indices can also be defined to account for disturbance effects. However, the steady-state operability analysis is actually a steady-state feasibility analysis in an *open-loop*. As such, it does not tell whether the AOS can be achieved using a feedback controller. Therefore, the OI analysis does not indicate the *closed-loop* operability properties, as Vinson and Georgakis themselves conceded [132].

Here we discuss an extension to the above steady-state operability analysis, which can be used to determine the steady-state operating points that are attainable in the closed loop, particularly using linear control. This approach [98] is based on the concept of passivity and the extended IMC framework presented in the previous section.

While most chemical processes are inherently nonlinear, the majority of control systems implemented in process industry are linear (particularly for regulating control) because they are simple to design, implement, operate and maintain. However, when the process is highly nonlinear, linear control may produce poor control performance, and even lead to system instability [89]. In principle, nonlinear processes can be controlled locally using a linear controller (although maybe only in a very small neighbourhood), provided their local linear approximations are controllable. Therefore, the real issue is which operating points are attainable (*i.e.*, offset-free controllable) via linear output feedback. Similar to [132], this approach requires only steady-state information on the process. In particular, the following two operating spaces of interest are discussed:

1. The *steady-state region of attraction* under linear feedback control [98], which defines the set of steady-state initial operating conditions in the input space from which the closed loop with linear control is guaranteed to converge *to the operating point of interest*.
2. The *steady-state output space achievable via linear feedback* [98], which defines the set of steady-state operating points in the output space to which the closed loop with linear control is guaranteed to converge *starting from the operating point of interest*.

The *steady-state region of attraction* is related to the solution of a regulation problem for the nonlinear closed loop, whilst the *achievable output space via linear feedback* is related to the solution of a servo problem. This section is based mainly on [98].

6.3.1 Steady-state Region of Attraction

Consider a nonlinear process G defined by the following nonlinear state-space model:

$$\begin{aligned}
\dot{x} &= f(x, u), \\
y &= g(x, u),
\end{aligned} \tag{6.22}$$

where $x \in \mathbb{R}^n$ are the state variables and $u, y \in \mathbb{R}^m$ are the process inputs and outputs, respectively. In addition, $f(\cdot) : \mathbb{R}^n \times \mathbb{R}^m \to \mathbb{R}^n$ and $g(\cdot) : \mathbb{R}^n \times \mathbb{R}^m \to \mathbb{R}^m$ are smooth vector valued functions. Assume that the model in (6.22) defines a steady-state nonlinear map $h(\cdot) : \mathbb{R}^m \to \mathbb{R}^m$ such that

$$y_{ss} = h(u_{ss}), \tag{6.23}$$

where $h(\cdot)$ is a smooth analytic vector valued function. Assume that DOS and AIS are nonempty and connected bounded subsets of $\mathbb{R}^m$; an operating point

$(u_{ss}^\star, y_{ss}^\star)$ that satisfies the nonlinear relation in (6.23) is said to be *feasible* if $u_{ss}^\star \in$ AIS and $y_{ss}^\star \in$ DOS.

Now we study conditions such that a *feasible* operating point $(u_{ss}^\star, y_{ss}^\star)$ can be attained using linear output feedback control. The proposed approach is based on the IMC framework discussed in Section 6.2.1. Consider the IMC structure shown in Figure 6.2, where G, $\tilde{G}$ and Q represent the process system, process model and IMC controller, respectively. Here we restrict the scope of the proposed conditions to processes G that are open-loop stable in the following sense:

Definition 6.5 (Asymptotic stability in a region [98]). *Consider a nonlinear process G defined by the nonlinear state-space model given in (6.22). It is said to be asymptotically stable in the region $X_0 \subset \mathbb{R}^n$ if every steady-state operating point (u_{ss}, x_{ss}, y_{ss}) that satisfies*

$$
\begin{aligned}
0 &= f(x_{ss}, u_{ss}), \\
y_{ss} &= g(x_{ss}, u_{ss}),
\end{aligned}
\tag{6.24}
$$

with $x_{ss} \in X_0$, is asymptotically stable for every initial condition x_0 in X_0.

For the nonlinear IMC structure, additional mild assumptions as given in Section B.4.1 are required to ensure the asymptotic stability of the nonlinear IMC closed-loop system. Assuming no model–plant mismatch, *i.e.*, $\tilde{G} = G$, $y(t) = GQr(t)$. In this case, the stability of the closed-loop system is determined by that of the IMC controller Q. A sufficient condition for offset-free control for a constant reference $r(t) = r_{ss}$ is that the steady-state mapping of the IMC controller Q be the right inverse of the process steady-state nonlinear map $h(\cdot)$ in (6.23).

As we discussed in Section 6.2.1, the IMC controller can be constructed by using two passive subsystems. If process G is strictly input passive, the IMC controller can be constructed by using a passive controller K with infinite gain in negative feedback with G, as shown in Figure 6.3. Here we provide a more general condition. Figure 6.6 shows how the IMC controller Q can be implemented as a feedback loop that contains a full dynamic nonlinear model $\tilde{G}$ of the process and a linear controller C with infinite gain at steady state. Without loss of generality, we assume that the controller C has the following form:

$$
C(s) = \bar{C}(s)\frac{\tilde{K}}{s},
\tag{6.25}
$$

where $\tilde{K} > 0$ is an $m \times m$ constant matrix and $\bar{C}(s)$ is a stable transfer function.

The set point $r_{ss}^\star$ is assumed such that the operating point $(u_{ss}^\star, y_{ss}^\star)$ with $y_{ss}^\star = r_{ss}^\star$ is feasible. The condition for steady-state attainability of $(u_{ss}^\star, y_{ss}^\star)$ can be derived from the asymptotic stability condition of the IMC controller Q, as shown in Figure 6.6, when an exogenous constant reference signal $r(t) = r_{ss}^\star$ is applied. The steady-state attainability condition is given as follows:

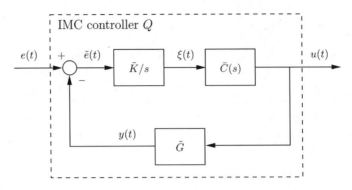

Fig. 6.6. Implementation of the IMC controller Q

Theorem 6.6 (Steady-state attainability via linear feedback control [98]). *Consider the closed-loop system shown in Figure 6.6. Assume that*

1. *The dynamics of the linear controller $\bar{C}(s)$ are described by the following state-space equations:*

$$\bar{C} \; : \; \begin{cases} \dot{z} = A\,z + B\,\xi \\ u = C\,z + D\,\xi, \end{cases} \qquad (6.26)$$

   (where $z \in \mathbb{R}^{n_z}$ and $\xi \in \mathbb{R}^m$) and matrix A in (6.26) is Hurwitz.
2. *The process G and the model $\tilde{G}$ are stable in the sense of Definition 6.5 and $\tilde{G} = G$.*
3. *The algebraic equation $f(x_{ss}, u_{ss}) = 0$ has a unique solution*

$$\bar{x}_{ss} = \psi(u_{ss}), \qquad (6.27)$$

   such that $\psi(\cdot)$ is C^2.
4. *The steady-state relation between ξ_{ss} and y_{ss} is given by*

$$y_{ss} = g(\bar{x}_{ss}, \bar{K}\xi_{ss}) = g\left(\psi(\bar{K}\xi_{ss}), \bar{K}\xi_{ss}\right) = h(\bar{K}\xi_{ss}) \triangleq \varphi(\xi_{ss}), \qquad (6.28)$$

   where $\bar{K} = -CA^{-1}B + D$ is the steady-state gain matrix of $\bar{C}(s)$ and $\varphi(\cdot)$ is a mapping from ξ_{ss} to y_{ss}.

Consider a constant reference $r(t) = r_{ss}^\star$ such that the corresponding operating point $(u_{ss}^\star, y_{ss}^\star)$ with $y_{ss}^\star = r_{ss}^\star$ is feasible. Let the integral action gain $\tilde{K}$ be

$$\tilde{K} = \varepsilon\hat{K}, \qquad (6.29)$$

where $\hat{K}^T = \hat{K} > 0$ and $\varepsilon > 0$ is a detuning coefficient. Assume that there exists a nonempty region $\Lambda_\xi \subset \mathbb{R}^m$ such that

$$[\xi - \xi_{ss}^\star]^T \left[h(\bar{K}\xi) - y_{ss}^\star\right] > 0, \quad \forall\, \xi \in \Lambda_\xi, \qquad (6.30)$$

where $y_{ss} = \varphi(\xi_{ss}) = h(\bar{K}\xi_{ss})$ is in a steady state. Then, there exists a (possibly small) $\varepsilon_0 > 0$ such that for all $0 < \varepsilon \leq \varepsilon_0$, the process equilibrium point $(u_{ss}^\star, y_{ss}^\star)$ is asymptotically stable if the closed-loop trajectory is such that $\xi(t) \in \Lambda_\xi \cup \{\xi_{ss}^\star\}$ for all $t \geq 0$.

The above theorem provides sufficient conditions for the asymptotic stability of a feasible equilibrium point $(u_{ss}^\star, y_{ss}^\star)$ using linear output feedback control. The key condition given in (6.30) is a *strictly input passivity condition* on the steady-state nonlinear mapping from ξ_{ss} to y_{ss} (including the nonlinear mapping $h(\cdot)$ and steady-state gain matrix $\bar{K}$ of the linear system $\bar{C}$). This result can be derived from the Passivity Theorem (Theorem 2.44). To provide a means to obtain the steady-state region of attraction, an alternative proof based on the Singular Perturbation Theorem was developed by the authors and their co-worker. The proof was first published in [98] and is included in Section B.4.2.

Because Theorem 6.6 is a steady-state condition, the gain of the integral action $\tilde{K} = \varepsilon \hat{K}$ ($0 < \varepsilon \leq \varepsilon_0$) can be arbitrarily small. Therefore, the dynamics of $\bar{C}(s)$ and the nonlinear model $\tilde{G}$ are irrelevant. While this theorem is about the controllability of a nonlinear process using linear feedback control, the analysis is based on the steady-state nonlinear model, beyond the studies on a linearised model around a feasible operating point. Now we can define the *steady-state region of attraction* mathematically:

Definition 6.7 (Steady-state region of attraction under linear feedback control [98]). *Consider a stable nonlinear process G and a feasible equilibrium point $(u_{ss}^\star, y_{ss}^\star)$. Denote the following ellipsoidal region $\Pi\left(\hat{K}, \gamma, u_{ss}^\star\right)$ in $\mathbb{R}^m$ centred at $u_{ss}^\star$ as*

$$\Pi_u\left(\hat{K}, \gamma, u_{ss}^\star\right) \triangleq \left\{ u \in \mathbb{R}^m \mid (u - u_{ss}^\star)^T \bar{K}^{-T} \hat{K}^{-1} \bar{K}^{-1} (u - u_{ss}^\star) \leq \gamma \right\}, \tag{6.31}$$

where $\gamma > 0$ is a scalar parameter. Then, the steady-state region of attraction under linear feedback control for the feasible operating point $(u_{ss}^\star, y_{ss}^\star)$ is given by

$$\Omega_u(u_{ss}^\star) \triangleq \max_{\hat{K}, \gamma} \mu\left[\Pi_u\left(\hat{K}, \gamma, u_{ss}^\star\right)\right],$$

subject to $\tag{6.32}$

$$\Pi_u\left(\hat{K}, \gamma, u_{ss}^\star\right) \subset \Lambda_u \cap AIS,$$

where Λ_u is the region that results from mapping the region Λ_ξ, where condition (6.30) in Theorem 6.6 holds, into the input space using the linear steady-state relation $u_{ss} = \bar{K}\xi_{ss}$. In addition, $\mu[\cdot]$ is a function that measures the hypervolume of the ellipsoid $\Pi\left(\hat{K}, \gamma, u_{ss}^\star\right)$, similar to that used in (6.20) and (6.21).

From the above definition, we can see that the steady-state region of attraction under linear feedback control $\Omega_u(u_{ss}^\star)$ is the largest ellipsoid completely inscribed in the *common* region of the AIS and the region where (6.30) holds. The actual size of $\Omega_u(u_{ss}^\star)$ depends on the chosen operating point $(u_{ss}^\star, y_{ss}^\star)$ because (6.30) is a static strictly input passivity condition at the equilibrium $(u_{ss}^\star, y_{ss}^\star)$ and thus Λ_u changes with different operating points $(u_{ss}^\star, y_{ss}^\star)$. The region of attraction $\Omega_u(u_{ss}^\star)$ is different from Λ_u because $\Omega_u(u_{ss}^\star)$ contains the set of steady-state initial operating conditions in the input space from which the closed loop with linear control is guaranteed to converge asymptotically to the operating point of interest $(u_{ss}^\star, y_{ss}^\star)$. A closed-loop trajectory that originates from any steady-state initial condition (u_{ss}, y_{ss}) such that $u_{ss} \in \Omega_u(u_{ss}^\star)$ will never leave the region $\Omega_u(u_{ss}^\star)$ for all $t \geq 0$ (*e.g.*, $\Omega_u(u_{ss}^\star)$ is a positively invariant set) and will converge to the feasible operating point $(u_{ss}^\star, y_{ss}^\star)$. The proof of the above statement is given in Section B.4.3 as Proposition B.9. Equation 6.32 also formulates the optimization problem that needs to be solved to determine the steady-state region of attraction. This is a standard problem in convex optimization theory [22].

The steady-state gain matrix $\bar{K}$ can be chosen such that $\Omega_u(u_{ss}^\star)$ is guaranteed to be nonempty. In this case, $\bar{K}$ serves as a scaling matrix such that (6.30) is satisfied in at least the neighbourhood of $(u_{ss}^\star, y_{ss}^\star)$. Assume that the Jacobian of the process steady-state nonlinear mapping evaluated at the equilibrium point $(u_{ss}^\star, y_{ss}^\star)$ is denoted as

$$J \triangleq \left. \frac{\partial h}{\partial u} \right|_{u=u_{ss}^\star}. \tag{6.33}$$

One natural choice of $\bar{K}$ is

$$\bar{K} = J^{-1}. \tag{6.34}$$

It can be proved that with (6.34), the steady-state region of attraction under linear feedback control $\Omega_u(u_{ss}^\star)$ is not empty [98].

The concept of the region of attraction can be illustrated by the following example from [98]:

Example 6.8 ([98]). Consider a two-input, two-output, stable nonlinear process whose input-output steady-state nonlinear mapping is given by the surfaces shown in Figure 6.7. Consider the following feasible equilibrium:

$$u_{ss}^\star = \begin{bmatrix} 2 \\ 4 \end{bmatrix}, \quad y_{ss}^\star = \begin{bmatrix} 17.63 \\ 24.69 \end{bmatrix}. \tag{6.35}$$

Assume that the AIS is the following square region:

$$\text{AIS} = \{(u_1, u_2) \mid -10 \leq u_1 \leq 10 \text{ and } -10 \leq u_2 \leq 10\}. \tag{6.36}$$

From Figure 6.7, it can be seen that there are points in the input space that exhibit input multiplicity (*i.e.*, there exist multiple different u_{ss} values that

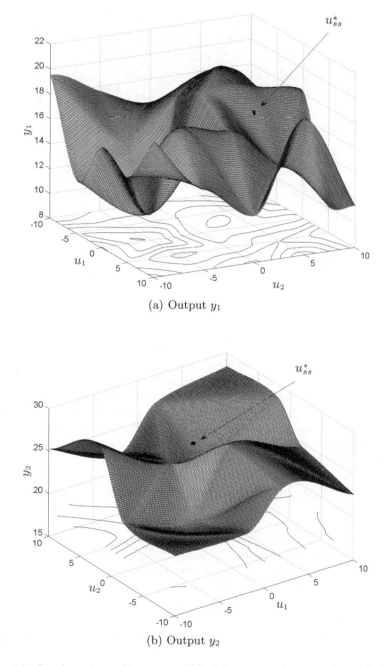

(a) Output y_1

(b) Output y_2

Fig. 6.7. Steady-state nonlinear map $h(\cdot)$ of the process in Example 6.8 [98]

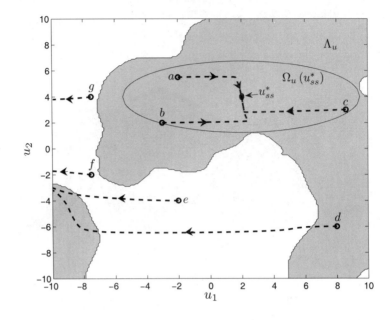

Fig. 6.8. Steady-state region of attraction and closed-loop trajectories [98]

map onto the same y_{ss}). The shaded region in Figure 6.8 is the region Λ_u in the input space in which (6.30) is satisfied. The filled circle inside Λ_u is the operating point $(u_{ss}^\star, y_{ss}^\star)$. The steady-state region of attraction under linear feedback control $\Omega_u(u_{ss}^\star)$ is bounded by the largest ellipse completely inscribed in Λ_u and in the AIS. The dashed curves in Figure 6.8 show the closed-loop trajectories obtained when the process G has steady-state initial conditions represented by the circles. It can be seen that every closed-loop trajectory with a steady-state initial condition inside $\Omega_u(u_{ss}^\star)$ (from points a, b or c) converges to the equilibrium $(u_{ss}^\star, y_{ss}^\star)$ using linear output feedback control. On the other hand, closed-loop trajectories with steady-state initial conditions outside region Λ_u (from points e, f and g) diverge and eventually move out of the input region. It is interesting to note that the closed-loop trajectory with steady-state initial condition inside region Λ_u but outside $\Omega_u(u_{ss}^\star)$ (point d) does not converge to the equilibrium $(u_{ss}^\star, y_{ss}^\star)$. This shows that region Λ_u, which satisfies the passivity condition (6.30) alone, does not represent the region of attraction.

6.3.2 Steady-state Output Space Achievable via Linear Feedback Control

Steady-state output space achievable via linear feedback control addresses a problem related to the region of attraction: given a feasible operating point

$(u_{ss}^{\star}, y_{ss}^{\star})$, what is the set of operating points the closed loop can switch to *starting from* $(u_{ss}^{\star}, y_{ss}^{\star})$ using *one* linear controller? This is a servo controllability problem and can be analysed by using the same IMC framework we used in Section 6.3.1, as shown in Figure 6.2. This time, we look at the conditions for which the IMC controller Q in Figure 6.3 is asymptotically stable when the reference signal $r(t)$ switches from $r(t) = y_{ss}^{\star}$ to a new value $r(t) = y_{ss}'$. Assume that system K is passive, $G_p = G = \tilde{G}$, and (u_{ss}', y_{ss}') is a feasible operating point. Define $\Delta u = u - u_{ss}'$, $\Delta y = Gu - y_{ss}'$, $\bar{G} : \Delta u \longmapsto \Delta y$. According to the Passivity Theorem, the sufficient condition for the IMC controller to be asymptotically stable (*i.e.*, $e = 0$ is an asymptotically stable equilibrium) is that system $\bar{G}$ is strictly input passive for *any* (u_{ss}', y_{ss}') in the region of interest. This is an incrementally strictly input passivity condition on process system G according to the following definition:

Definition 6.9 (Incremental input passivity [32]). *Let* $H : \mathcal{L}_{2e}^m \to \mathcal{L}_{2e}^m$. *System H is said to be incrementally input passive if*

$$\langle Hu - Hu', u - u' \rangle_T \geq 0, \quad \forall\ u,\ u' \in \mathcal{L}_{2e}^m. \tag{6.37}$$

System H is said to be incrementally strictly input passive if there exists a constant ν such that

$$\langle Hu - Hu', u - u' \rangle_T \geq \nu \|u - u'\|_T^2, \quad \forall\ u,\ u' \in \mathcal{L}_{2e}^m. \tag{6.38}$$

Now we generalize the above result with the linear control shown in Figure 6.6:

Theorem 6.10 (Steady-state attainability via linear feedback control for step changes in reference). *Consider the closed-loop system shown in Figure 6.6. Assume that $\hat{K} > 0$ and the feasible operating point $(u_{ss}^{\star}, y_{ss}^{\star})$ is attainable using linear feedback control for all $0 < \varepsilon \leq \varepsilon_0$, based on the assumptions and conditions outlined in Theorem 6.6. Define a static mapping $\varphi(\xi_{ss}) = h(\bar{K}\xi_{ss})$, where $u_{ss} = \bar{K}\xi_{ss}$. Consider a step change in the reference signal $r(t)$ such that*

$$r(t) = \begin{cases} y_{ss}^{\star}, & t < 0 \\ y_{ss}', & t \geq 0, \end{cases} \tag{6.39}$$

and assume that the new operating point (u_{ss}', y_{ss}') is also feasible. Assume that $h(\cdot)$ satisfies the following Lipschitz continuity condition:

$$\|h(u + \Delta u) - h(u)\| \leq \gamma \|\Delta u\|, \tag{6.40}$$

for all u in the AIS and $0 \leq \gamma < \infty$. Suppose that there exists a nonempty region $\Theta_u \subset AIS \subset \mathbb{R}^m$ such that

1. The following condition is satisfied:

$$\frac{\partial h}{\partial u_{ss}} \bar{K} > 0, \quad \forall\ u_{ss} \in \Theta_u, \tag{6.41}$$

or

2. *The steady-state mapping $\varphi(\cdot)$ is incrementally strictly input passive for any $\xi_{ss} = \bar{K}^{-1} u_{ss}$, $\forall u_{ss} \in \Theta_u$.*

Then the new operating point (u'_{ss}, y'_{ss}) is asymptotically stable if the closed-loop trajectory is such that $u(t) \in \Theta_u$ for all $t \geq 0$.

The above theorem is an extended version of Theorem 2.2 in [98]. While the incremental input passivity condition is more intuitive and reveals the link between controllability and passivity, it is very hard to check numerically. Condition 1 is much easier to verify and thus better used in controllability analysis. The proof of Condition 1, which is slightly different from that given in [98] is presented in Section B.4.4. The equivalence of Conditions 1 and 2 was proved in [97]. The *steady-state output space achievable via linear feedback control* is defined as follows:

Definition 6.11 (Steady-state output space achievable via linear feedback control). *Consider a stable nonlinear process G and a feasible equilibrium point $(u^\star_{ss}, y^\star_{ss})$. Denote the following ellipsoidal region $\Pi_y\left(\hat{K}, \gamma, y_{ss}\right)$ in $\mathbb{R}^m$ centred at y_{ss} as*

$$\Pi_y\left(\hat{K}, \gamma, y_{ss}\right) \triangleq \left\{ y \in \mathbb{R}^m \mid (y_{ss} - y)^T \hat{K}(y_{ss} - y) \leq \gamma \right\}. \qquad (6.42)$$

A steady-state operating point y_{ss} is achievable from $(u^\star_{ss}, y^\star_{ss})$ via linear feedback control if there exists a $\hat{K}$ and γ such that the ellipsoidal region $\Pi_y\left(\hat{K}, \gamma, y_{ss}\right)$ is completely inscribed in Θ_y and covers $y^\star_{ss}$, that is,

$$y^\star_{ss} \in \Pi_y\left(\hat{K}, \gamma, y_{ss}\right) \subset \Theta_y, \qquad (6.43)$$

where Θ_y is the region in the output space that results from mapping the region Θ_u where (6.41) holds, using the nonlinear steady-state relation $y_{ss} = h(u_{ss})$. The steady-state output space achievable via linear feedback control for the feasible operating point $(u^\star_{ss}, y^\star_{ss})$ is the set of all y_{ss} which are achievable from $(u^\star_{ss}, y^\star_{ss})$ with a single constant $\hat{K}$ but possibly different γ.

The above definition is an extension of the achievable output space defined in [98]. This is illustrated by Figure 6.9 for an arbitrary region Θ_y. The steady-state output space achievable via linear feedback control $\Omega_y(y^\star_{ss})$ contains the set of steady-state operating points in the output space to which the closed loop with linear control is guaranteed to converge starting from the operating point of interest $(u^\star_{ss}, y^\star_{ss})$.

Now let us study the steady-state output space achievable of the process system in Example 6.8:

Example 6.12 ([98]). Consider stable nonlinear process G described in Example 6.8 with the same available input space (AIS) and the feasible operating

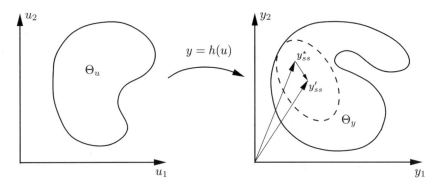

Fig. 6.9. Geometric interpretation of output attainability

point $(u_{ss}^\star, y_{ss}^\star)$ in (6.35). Assume that $\bar{K}$ and $\hat{K}$ are those selected in Example 6.8. First, we find the region Θ_u such that the condition $\frac{\partial h}{\partial u}\bar{K} > 0$ is satisfied. The region Θ_u is then mapped into the output space using the nonlinear steady-state relation $y_{ss} = h(u_{ss})$. The result is the irregular shaded region Θ_y shown in Figure 6.10. The points y_{ss} that satisfy (6.43) are obtained numerically. The resulting output space achievable via linear feedback control $\Omega_y(y_{ss}^\star)$ is given by the region shown in Figure 6.10 around the steady-state output $y_{ss}^\star = [17.63, \ 24.69]^T$ which is represented by a filled circle. A magnified image of $\Omega_y(y_{ss}^\star)$ is shown in Figure 6.11. The dashed lines in Figure 6.11 show the simulated closed-loop trajectories of the process output when the reference signal $r(t)$ changes from $y_{ss}^\star$ to the value y_{ss}' marked with squares. Observe that since $y_{ss}' \in \Omega_y(y_{ss}^\star)$, the closed-loop trajectories are guaranteed to converge to the new operating point. Unlike the steady-state region of attraction for linear feedback control $\Omega_u(u_{ss}^\star)$, the steady-state output space achievable $\Omega_y(y_{ss}^\star)$ is not positively invariant for $y(t)$. This is clearly shown in Figure 6.11. The closed-loop output trajectory converges toward the steady-state point in the upper-left corner, exits and then reenters the region $\Omega_y(y_{ss}^\star)$. The trajectory remains inside the region Θ_y, where (6.41) holds.

6.3.3 Steady-state Attainability by Nonlinear Control

From the analysis in the previous subsections, we can obtain more insights into nonlinearity and linear controllability. A variety of nonlinearity measures have been proposed to quantify the degree of nonlinearity of the (open-loop) process, *e.g.*, [33, 48, 118]. The following statement by Eker and Nikolau [36] represents the common belief: "The premise of these approaches is that if a nonlinear open-loop system is far from a linear one, then linear control will, most probably, be inadequate for the closed loop." However, it has been recently discovered that the degree of nonlinearity of the open-loop process may not always be related to poor performance with a linear controller [36, 89].

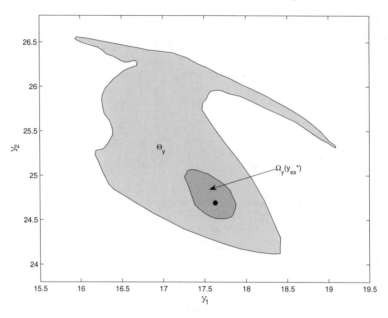

Fig. 6.10. Regions Θ_y and $\Omega_y(y_{ss}^\star)$ [98]

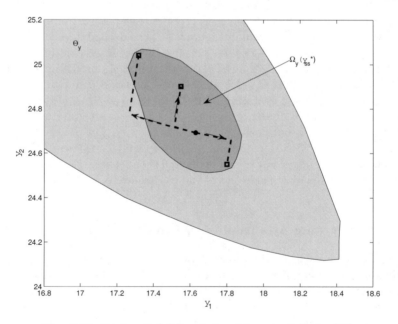

Fig. 6.11. Region $\Omega_y(y_{ss}^\star)$ and closed-loop trajectories [98]

This is precisely what we have seen from the passivity-based controllability analysis: A highly nonlinear process may not necessarily be difficult to control if the operating region satisfies the passivity condition (6.30). Example 6.8 shows that a highly nonlinear process may satisfy (6.30) in a large region Λ_u. From the shape of the surfaces shown in Figure 6.7, we can see that a single linearised model of the process would have serious difficulties in describing process G over the entire region Λ_u. Despite this, the above analysis shows that linear control can guarantee offset-free performance if the steady-state initial condition is inside the steady-state region of attraction $\Omega_u(u_{ss}^\star)$.

Readers may have noticed that in the steady-state attainability analysis approach presented in this section, we do not approximate the steady-state process model $y_{ss} = h(u_{ss})$. Instead, we find the regions Θ_u and Θ_y in which the mapping $h(u_{ss})$ is strictly input passive (for the region of attraction) and incrementally strictly input passive (for the output space achievable) because we need to find the steady-state attainability by *linear* control: the final controller in the extended IMC structure shown in Figure 6.3 is guaranteed to be linear if $G_p = \tilde{G}$. In this case, process G is effectively controlled in closed loop by the linear controller $C(s)$ in (6.25). If we use a passivated model $G_p \neq \tilde{G}$, then the final controller will be $C(s)$ with the negative feedback of $G_{ff} = G_p - \tilde{G}$, leading to a nonlinear final controller. Because only steady-state attainability is concerned, we just need to passivate the steady-state mapping $h(\cdot)$. In this case, offset-free control may not be achievable in the region where the passive approximation $h_p(\cdot)$ differs from $h(\cdot)$. The problem is to find a mapping $h_p(\cdot)$ such that the volume of the input space in which $h_p(u_{ss}) = h(u_{ss})$ is maximized. By implementing $h_p(\cdot)$, the regions of steady-state attainability will be larger because both Θ_u and Θ_y will be larger. This explains why a nonlinear controller can work effectively over a larger range of operating conditions than a linear controller.

One numerical approach to obtain the passive approximation is to assume that $h_p(\cdot)$ can be represented as the linear combination of nonlinear base functions $\phi(\cdot)$, *e.g.*,

$$h_p(u_{ss}) = \sum_{i=1}^{k} w_i \phi_i(u_{ss}). \tag{6.44}$$

In this case, the passivation problem is converted into an optimization problem with the decision variables of w_i to minimize the difference between $h_p(\cdot)$ and $h(\cdot)$, subject to the passivity condition on $h_p(\cdot)$. This problem can be solved numerically. Spline functions are good candidates for the nonlinear base function $\phi(\cdot)$ because they are smooth and very flexible. A numerical approach to passivation using spline functions can be found in [28].

A steady-state mapping function $y_{ss} = h(u_{ss})$ with input multiplicity implies operational difficulties, because no unique solution u_{ss} is available for a desired output value y_{ss}. The steady-state attainability condition (6.30) implies that there is no input multiplicity in the neighbourhood of $(u_{ss}^\star, y_{ss}^\star)$. The attainability condition given in Theorem 6.10 for the achievable output space requires more. The incrementally strict input passivity condition implies that there is no operating point that exhibits input multiplicity in the entire region Θ_u.

6.3.4 Numerical Procedure

Although the discussions in Sections 6.3.1 and 6.3.2 are based on the steady-state mapping $h(\cdot)$, the attainability analysis is often more conveniently performed based on the numerical values of steady-state operating points (u_{ss}, y_{ss}) rather than an explicit nonlinear function $h(\cdot)$. In this case, a set of operating conditions (u_{ss}, y_{ss}) with sufficient resolution is required. The steady-state operating conditions can be obtained by using process simulation software packages, $e.g.$, ASPEN Plus®. The numerical procedure for the steady-state attainability analysis is as follows:

Procedure 6.13 (Steady-state attainability analysis)

1. *Obtain the steady-state operating conditions using a process simulation package*
   a) *Build the process flowsheet using a process simulation package.*
   b) *Define the AIS based on the process design.*
   c) *Generate a number of equally spaced points for each input variable in the AIS.*
   d) *Derive the process steady-state output conditions by evaluating the process outputs for every input combination inside the AIS. This can be done easily using a sensitivity analysis tool in a process simulation package.*
   e) *Export the process output results to be used in the next step.*
2. *Study the steady-state region of attraction under linear feedback control $\Omega_u(u_{ss}^\star)$:*
   a) *Select an operating point of interest $(u_{ss}^\star, y_{ss}^\star)$. Calculate the Jacobian matrix J of the steady-state nonlinear map at the operating point $(u_{ss}^\star, y_{ss}^\star)$ and corresponding $\bar{K}$ as in (6.33) and (6.34). Calculate $\xi_{ss} = \bar{K}^{-1} u_{ss}$.*
   b) *Calculate the region Λ_u corresponding to all $\xi_{ss} \in \Lambda_\xi$, where Λ_ξ is the region defined in (6.30).*
   c) *Calculate $\Omega_u(u_{ss}^\star)$ by solving the optimization problem in (6.32).*
3. *Study the steady-state output space achievable via linear feedback control $\Omega_y(y_{ss}^\star)$:*
   a) *Calculate the region $\Theta_u \subset AIS$ defined in (6.41).*
   b) *Find the set of corresponding steady-state output values Θ_y using a process simulation package for $u_{ss} \in \Theta_u$.*
   c) *Calculate the $\Omega_y(y_{ss}^\star)$ by solving the problem given in (6.43).*

6.3.5 Case Study of a High-purity Distillation Column

In this subsection, we illustrate the steady-state attainability analysis using a high-purity distillation column [106]. Distillation columns have been well studied in the context of process control. However, most of the distillation systems considered in the literature are low to moderate purity separation systems. High-purity distillation columns are highly nonlinear and are known to be difficult to control [128].

A high-purity distillation column for methanol-water separation is considered. This distillation column system is similar to that studied by Chiang and Luyben [30]. The feed stream consisting of 50 mol% methanol is fed into the column at Tray 67 of a 99 tray column. This column is operated at 1.172 bar. The distillate product is expected to have a high-purity of 99.9 mol% methanol while the bottom product will be a water stream containing only 0.1 mol% methanol. This distillation column is designed by using a rigorous tray to tray calculation method. The calculation is carried out by using the RadFrac^{TM} model in the ASPEN Plus$^{\circledR}$ model library. The steady-state design specifications of the distillation column are summarized in Table 6.2.

Table 6.2. Design specifications of high-purity distillation column

Specification	Value
Feed rate (kmol/h)	2300
Feed temperature ($^{\circ}$C)	57.22
Feed composition	0.5
Distillate rate (kmol/h)	1150
Distillate temperature ($^{\circ}$C)	68.30
Distillate composition	0.999
Bottom rate (kmol/h)	1150
Bottom temperature ($^{\circ}$C)	103.96
Bottom composition	0.001
Operating pressure (bar)	1.172
Number of trays	99
Feed tray location (top tray= 1)	67
Tray efficiency	0.75
Reflux ratio	0.864
Reboiler heat duty (MMkcal/h)	19.286

The manipulated variables are the distillate rate (u_1) and the reboiler heat duty (u_2). The AIS is defined by $1050 \leq u_1 \leq 1250$ and $17 \leq u_2 \leq 21$ (both in kmol/h). By varying the values of one input variable over the values of the other manipulated variable, the corresponding output variables, namely, the water content in the distillate product y_1 and the methanol content in the bottom product y_2 (both in ppm) can be generated. The calculations were performed by using the sensitivity analysis tool in ASPEN Plus$^{\circledR}$. A grid of

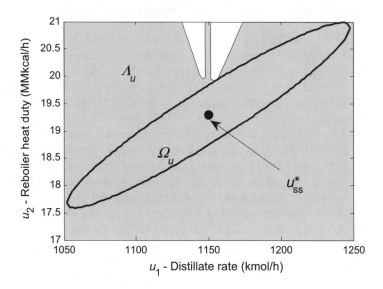

Fig. 6.12. Steady-state region of attraction of the distillation column

101×101 evenly spaced points defined on the input space is used to generate a total of corresponding 10201 output points.

Consider a feasible steady-state operating point of $u_{ss}^{\star} = [1150, 19.286]^{T}$ and $y_{ss}^{\star} = [1000, 1000]^{T}$. Following Procedure 6.13, both the steady-state region of attraction $\Omega_u (u_{ss}^{\star})$ and achievable output space $\Omega_y (y_{ss}^{\star})$ are obtained, as shown in Figures 6.12 and 6.13 respectively. From Figure 6.12, it can be seen that Λ_u covers a large area of the AIS. However, the region of attraction, $\Omega_u (u_{ss}^{\star})$, is much smaller. This means that only a fairly small portion of the initial operating points in the AIS can be driven toward the operating point of interest $(u_{ss}^{\star}, y_{ss}^{\star})$ in a closed loop by using linear control. For example, the initial operating point with $u_{ss} = [1100, 18.5]^{T}$ can be driven and guaranteed to converge to $(u_{ss}^{\star}, y_{ss}^{\star})$. On the other hand, there is no guarantee that the initial operating point with $u_{ss} = [1200, 18.5]^{T}$ which lies outside $\Omega_u (u_{ss}^{\star})$ but inside Λ_u, can be driven to $(u_{ss}^{\star}, y_{ss}^{\star})$. Therefore, if a linear controller is intended to be implemented, the startup operation is possible only from within $\Omega_u (u_{ss}^{\star})$. The achievable output space $\Omega_y (y_{ss}^{\star})$ covers an even smaller portion of the AOS. This indicates that the flexibility of changing the operating point of interest to a new operating point is very limited. This again confirms the difficulties of controlling this highly nonlinear process by using a linear controller.

These results provide a more realistic measure of the process operability compared to the OI analysis [132]. If the DOS is defined by $10 \leq y_1 \leq 10000$ and $10 \leq y_2 \leq 10000$, then OI= 1 because every desired steady-state output value in the DOS can be achieved by open-loop mapping of the steady-state

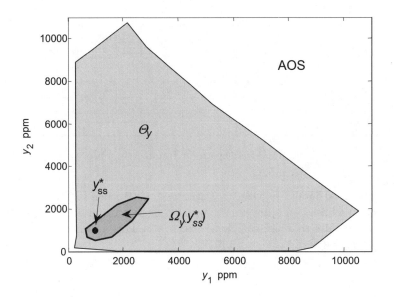

Fig. 6.13. Steady-state output space achievable of the distillation column

input within the AIS. However, the steady-state attainability analysis shows
that only a small operating region can be obtained via linear feedback control.

6.4 Dynamic Controllability Analysis for Nonlinear Processes

As we mentioned in the previous section, dynamic controllability analysis for
general nonlinear process systems is usually very difficult. Here we present a
controllability analysis approach based on approximate nonlinear models. This
section is based mainly on the recent development by the authors and their
co-worker which was reported in [97]. The central idea is to approximate the
nonlinear process model using models with static nonlinearity and linear dynamics. Studies have shown that in many cases, such approximate models can
describe nonlinear processes with sufficient accuracy [37, 74]. Linear approximations of the process dynamics can be obtained from process flowsheet data
using, for example, the approach proposed by Lewin and co-workers [109, 134].
Therefore, based only on process flowsheet data for a nonlinear process, one
may derive an approximate model in the form of a series interconnection of
static nonlinearities and linear dynamics:

$$y = N_o G_l N_i u, \tag{6.45}$$

where $N_i : \mathbb{R}^m \to \mathbb{R}^m$ and $N_o : \mathbb{R}^m \to \mathbb{R}^m$ are input and output static nonlinearities, respectively, and G_l is a linear multivariable model. The above

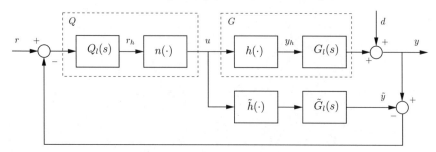

Fig. 6.14. IMC framework for Hammerstein processes

block-structured model is called a Hammerstein–Wiener system. Controllability analysis based on approximate process models is quite manageable and often gives results that are sufficiently accurate. This approach is built on a further extended IMC framework [97].

Assume that process model G is represented by a Hammerstein system:

$$y_h = h(u), \tag{6.46}$$

$$y = G_l y_h, \tag{6.47}$$

where $h(\cdot)$ is a static nonlinear function. As shown in Figure 6.14, the IMC controller will be a Wiener system, consisting of a static nonlinear function $n(\cdot)$ and linear dynamic system Q_l, which are ideally the inverses of $h(\cdot)$ and G_l, respectively. If $n(\cdot) = h^{-1}(\cdot)$, then,

$$r_h = n^{-1}(u) = h(u). \tag{6.48}$$

The invertibility of the static nonlinear function can be determined by the following theorem:

Theorem 6.14 ([32]). *Consider a nonlinear mapping $y_h = h(u) : \Omega_u \subset \mathbb{R}^m \to \Omega_{y_h} \subset \mathbb{R}^m$. If there exists a nonsingular constant matrix $\bar{K}$ such that $h(\bar{K}u)$ is incrementally input passive and the Lipschitz continuity condition given in (6.40) is satisfied for all $u \in \Omega_u$ and $u + \Delta u \in \Omega_u$ with $0 \le \gamma < \infty$, then, $h^{-1}(\cdot)$ exists and is a well-defined map from Ω_{y_h} to Ω_u.*

The above condition is equivalent to Condition 2 of Theorem 6.10. Practically, we can use (6.41) with $\bar{K} = \left(\frac{\partial h}{\partial u}\big|_{u=u^\star}\right)$ to determine the input region Θ_u in which the mapping $h(\cdot)$ is incrementally strictly input passive ($u^\star$ is an operating point that belongs to the DIS). If $n(\cdot) = h^{-1}(\cdot)$, then the output of the linear system Q_l should be confined in the following region:

$$\Theta_{rh} \triangleq h(\Theta_u) = \Theta_y. \tag{6.49}$$

Note that because the final controller is nonlinear, the output space of the static nonlinearity does not need to be confined in the steady-state output space achievable via linear feedback control given in (6.43). Now the achievable performance of the process can be evaluated by the best dynamic control performance that the linear IMC controller Q_l can deliver, subject to the constraint: $r_h \in \Theta_{rh}$. This constraint on $r_h(t)$ represents the influence that the static nonlinearity $h\{\cdot\}$ has on the operability of the overall nonlinear process G. The smaller the invertibility region Θ_{r_h} the stronger the limitation imposed by the static nonlinearity.

One possible way to obtain the dynamic performance achievable is to solve a constrained nonlinear optimization problem in the time domain, similar to the optimization approach adopted in the simultaneous process and control design methodology (*e.g.*, [103]). However, this approach requires huge computational effort and is limited to simple process systems. An easier approach is to replace the time-domain "hard" constraint $r_h(t) \in \Theta_{r_h}$ by a related (though generally more conservative) constraint on the system norm of the IMC controller Q_l and evaluate the performance of an optimal IMC controller that is designed subject to the system norm constraint.

Consider a regulating control problem with output disturbance d, as shown in Figure 6.14. To assist the controllability analysis, the input and output variables should be normalised componentwise. For example, the normalised output $\tilde{y} = [\tilde{y}_1, \tilde{y}_2, \ldots, \tilde{y}_m]^T$ is obtained for $y = [y_1, y_2, \ldots, y_m]^T$ by the following operation:

$$\tilde{y}_i = y_i / y_{i\,\mathrm{max}}, \tag{6.50}$$

where $y_{i\,\mathrm{max}}$ is the largest possible value of y_i. Similarly, we can obtain the normalised variables $\tilde{r}, \tilde{r}_h, \tilde{u}$ and $\tilde{d}$. If $r(t) = y^\star$, $\tilde{r}(t) = -\tilde{d}(t)$. A natural choice is bound by both the disturbance $\tilde{d}$ and $\tilde{r}_h$ using their ∞-norm, which is the largest possible value of the all elements of the vector signal for all times. For example, for $\tilde{d}(t) = \left[\tilde{d}_1(t), \ldots, \tilde{d}_m(t)\right]^T$,

$$\|\tilde{d}\|_\infty \triangleq \max_\tau \left(\max_i \left| \tilde{d}_i(\tau) \right| \right). \tag{6.51}$$

Then the system norm of Q_l can be bounded by its $\mathcal{L}_1$-norm $\|q_l(t)\|_1$:

$$\|q_l\|_1 \triangleq \int_{-\infty}^{\infty} q_l(t)\, dt = \sup_{\tilde{d}} \frac{\|\tilde{r}_h\|_\infty}{\|\tilde{d}\|_\infty}, \tag{6.52}$$

where $q_l(t)$ is the impulse response of the linear system Q_l. Unfortunately, this may lead to extremely conservative system norm bounds because the ∞-norm of a signal does not well represent the impact the signal can have on a dynamic system. For example, rectangular pulse input signals with the same height but different widths will have the same ∞-norm but can lead to very different system outputs. A better way to quantify the disturbance is to use its 2–norm, related to the energy level:

$$\|\tilde{d}\|_2 \triangleq \sqrt{\int_0^\infty \tilde{d}(t)^T \tilde{d}(t)\, dt} = \upsilon < \infty. \tag{6.53}$$

We then bound $\tilde{r}_h(t)$ with the $\infty - 2$ norm, defined as

$$\|\tilde{r}_h\|_{\infty-2} \triangleq \sup_{\tau} \|\tilde{r}_h(\tau)\|, \tag{6.54}$$

which is the largest possible Euclidean vector norm of $\tilde{r}_h(t)$ for all times (remember we are dealing with multivariable systems), where

$$\|\tilde{r}_h(\tau)\| = \sqrt{\tilde{r}_h(\tau)^T \tilde{r}_h(\tau)}. \tag{6.55}$$

Now we can approximate the constraints on $\tilde{r}_h(t)$ using the following generalized $\mathcal{H}_2$-norm of Q_l:

$$\|Q_l\|_g = \sup_{\tilde{d}} \frac{\|\tilde{r}_h\|_{\infty-2}}{\|\tilde{d}\|_2}. \tag{6.56}$$

First introduced by Wilson [141], the generalized $\mathcal{H}_2$-norm constraint can be easily incorporated in control design. Assuming that system $Q_l(s)$ has the state-space representation of (A_q, B_q, C_q, D_q), its generalized $\mathcal{H}_2$-norm is less than γ if and only if there exists a symmetrical matrix P such that the following LMIs are satisfied [107]:

$$\begin{bmatrix} A_q^T P + P A_q & P B_q \\ B_q^T P & -I \end{bmatrix} < 0,$$

$$\begin{bmatrix} P & C_q^T \\ C_q & \gamma^2 I \end{bmatrix} > 0, \tag{6.57}$$

$$D_q = 0.$$

The sufficient condition for $\|\tilde{r}_h\|_{\infty-2} < \alpha$ for all $\tilde{d}(t)$ subject to $\|\tilde{d}\|_2 < v$ is that

$$\|Q_l\|_g < \alpha v^{-1}. \tag{6.58}$$

When $\tilde{r}_h(t) \in \Theta_{r_h}$, $n(\cdot)$ inverts $h(\cdot)$, and thus the dynamic performance of the linear IMC controller can be quantified by the $\infty-$norm of the sensitivity function:

$$\|S\|_\infty = \|I - G_l Q_l\|_\infty = \sup_{\tilde{d}} \frac{\|\tilde{e}\|_2}{\|\tilde{d}\|_2}. \tag{6.59}$$

Now the dynamic controllability assessment problem is formulated as the following problem:

Problem 6.15 ([97]).

$$\min_{Q_l(s)} \|I - G_p Q_p\|_\infty, \tag{6.60}$$

subject to

$$\|Q_l\|_g < \alpha v^{-1}. \tag{6.61}$$

By using the bounded-real lemma (Lemma 3.17) and (6.57), the constraints in the above optimization problem can be represented using matrix inequalities, which in turn can be converted in LMIs. The sensitivity function with the smallest ∞-norm can be found by solving the LMI problem. More details can be found in [97].

6.5 Summary and Discussion

In this chapter, we presented some recent developments of process input-output controllability analysis based on the concept of passivity, including the link between passivity, process invertibility and controllability. By extending the IMC structure, we can quantify the achievable control performance based on the passive approximation of process systems. Based on this framework, we introduced steady-state attainability analysis. Using only the process steady-state operating conditions which are often available in the process design stage, the steady-state attainability analysis is a useful tool for examining the impact of process nonlinearity on its controllability. Also based on the extended IMC structure, we discussed dynamic controllability analysis methods for nonlinear process systems using approximate models in the Hammerstein form.

The analytical approaches presented in this chapter are valid for input-output stable processes. For unstable processes, the controllability should be determined based on both IFP and OFP indices. The OFP index indicates the minimum gain of the controller required to stabilize an unstable process. However, IFP and OFP indices are not independent from each other. It is also very hard to obtain dynamic IFP and OFP indices, particularly for nonlinear systems. An effective controllability analysis approach for nonlinear unstable processes is still required.

Passivity-based controllability analysis is an emerging research area. New developments are still evolving, among which the following trends are particularly promising:

- *Controllability analysis based on the link between thermodynamics and passivity.* Pioneered by Ydstie and co-workers, the links between certain process systems which obey the laws of thermodynamics and their passivity have been recently revealed [2, 3, 143, 144] (which are detailed in Chapter 7). It is possible to determine process controllability from the irreversible thermodynamics of process systems. This may lead to heuristic design rules for better process controllability for chemical engineers.
- *Controllability analysis for process system networks.* This is a rather interesting topic, because often the coupling of process units has a major effect on plantwide controllability. Being an input-output property, passivity is particularly useful in determining the stability of a network of interconnected processes. It is possible to perform controllability analysis for complex process networks based on the passivity of each process unit

and the topology of the interconnection. This will be useful in pinpointing the sources of controllability problems caused by process interactions.

7

Process Control Based on Physically Inherent Passivity

by K.M. Hangos and G. Szederkényi

General passivity-based control is difficult without physical insights. Even at the origin of passivity-based control, the physical analogies of stored and retrievable energy of mechanical systems have been used fruitfully. Later, the notions and techniques to apply the fundamental physical description to design controllers emerged particularly in the area of mechanical systems and robotics [115, 130]. Inspired by this fruitful connection, this chapter is devoted to the thermodynamic foundation of process control.

Process systems can be seen as a class of systems that obey the laws of thermodynamics. Therefore, it is natural to use thermodynamics as the underlying physical theory for constructing passivity-based controllers instead of theoretical mechanics.

The idea of investigating the dissipation and passivity of process systems based on thermodynamic principles was introduced in the 1990s [2, 38, 100, 144], where its implications for controller design have also been explored. The approach has also been applied to networks of process systems, *i.e.*, to composite process systems with several balance volumes in [52].

Following this starting point, there is a wide and growing literature in the field of connections between thermodynamics, variational calculus and the theory of Hamiltonian systems. Ydstie [143] offers a recent survey of related papers. The principles of constructing a Hamiltonian system model for process systems can be found in [53].

The aim of this chapter is to explore the deeply rooted physical fundamentals of passivity-based process control that lie in thermodynamics and in first principle modelling of process systems. Because of space limitations, we could only aim at a brief introduction to the subject. The interested reader is encouraged to study the references cited for more details.

7.1 Thermodynamic Variables and the Laws of Thermodynamics

Thermodynamics is taught to process and mechanical engineers, physicists and chemists as a basic first year subject, yet it constitutes one of the fundamentals of process modelling and control. It "$\cdots$ does not predict specific numerical values for observable quantities. Instead, thermodynamics sets limits (*inequalities*) on permissible physical processes $\cdots$" [25]. Therefore, it is essential from the viewpoint of passivity-based process control.

Thermodynamics influences control design in various ways: it determines the number and kind of state variables in our models and provides physically motivated storage or Lyapunov function candidates, among others.

Classical thermodynamics deals with closed systems in equilibrium or near equilibrium, but process systems are usually open and in a transient state. Therefore, the basic principles of nonequilibrium thermodynamics [72] are also of interest for process control.

7.1.1 Extensive Variables, Entropy, Intensive Variables

A simple axiomatic approach to thermodynamics is to formulate the fundamental statements in the form of *postulates* and derive their consequences. The approach taken here follows the description by Callen [25].

For the sake of simplicity, we consider the simplest case, when the considered process system consists of homogeneous closed domains of space called *simple regions*.

Extensive variables

According to *Postulate I* of thermodynamics, the equilibrium states of a simple region are characterized completely by the internal energy U, the volume V, and the mole numbers N_1, $\cdots$, N_K of the chemical components forming the canonical set of extensive variables:

$$U, \ V, \ N_1, \ldots, N_K. \tag{7.1}$$

This implies that every other extensive quantity is a function of the above set.

Note that extensive quantities are strictly additive when joining two regions.

Entropy

An additional extensive variable, so-called *entropy* (denoted as S), is used to characterize an equilibrium state that eventually results after the removal of constraints in a composite system of two or more simple regions.

Postulate II states that the values of the extensive quantities in the absence of an internal constraint are those that maximize the entropy over the manifold of constrained equilibrium states.

Postulate III furthermore states that entropy is extensive, continuous and differentiable and is a monotonically increasing function of the energy.

From *Postulate I* and *Postulate III*, it follows that

$$S = S(U, V, N_1, \ldots, N_K), \quad \left(\frac{\partial S}{\partial U}\right)_{V, N_1, \ldots, N_K} > 0. \tag{7.2}$$

The above equation is called the *entropic fundamental equation*.

In addition to the monotonically increasing property of entropy as a function of internal energy, entropy is a *concave function* in any of its independent variables, that is, for example,

$$\frac{S(U_1, V, N_1, \ldots, N_K) + S(U_2, V, N_1, \ldots, N_K)}{2} \leq S(\frac{U_1 + U_2}{2}, V, N_1, \ldots, N_K). \tag{7.3}$$

This means that the average entropy of two regions that differ only in their internal energy is smaller than that of the same region with an average internal energy. This is a form of *Postulate II* implying that the internal energy would "equilibrate" when joining regions.

Alternatively to (7.2), we can consider energy U as an independent variable depending on the others, giving rise to the so-called *fundamental equation*:

$$U = U(S, V, N_1, \ldots, N_K). \tag{7.4}$$

Intensive variables

Because of our interest in dynamics and in the associated changes in extensive quantities, the differential form of the fundamental equation (7.4) is computed in the form:

$$dU = \left(\frac{\partial U}{\partial S}\right)_{V, N_1, \ldots, N_K} dS + \left(\frac{\partial U}{\partial V}\right)_{S, N_1, \ldots, N_K} dV + \sum_{j=1}^{K} \left(\frac{\partial U}{\partial N_j}\right)_{S, V, \ldots, N_K} dN_j. \tag{7.5}$$

The various partial derivatives are called *intensive quantities* and are conventionally denoted as follows:

$\left(\frac{\partial U}{\partial S}\right)_{V, N_1, \ldots, N_K}$ is the temperature T,

$\left(\frac{\partial U}{\partial V}\right)_{S, N_1, \ldots, N_K}$ is the negative pressure $-p$ and

$\left(\frac{\partial U}{\partial N_j}\right)_{S, V, \ldots, N_K}$ is the chemical potential μ_j of component j.

Unlike extensive variables, intensive variables are not additive when joining two or more regions, but they *equilibrate*. The significance of intensive variables is explained by the fact that they are the usual measurable quantities in process systems. For later use, we collect the *canonical set of engineering intensive variables* for a region in a vector:

$$\chi_0 = p \ , \ \chi_1 = T \ , \ \chi_j = c_{j-1}, \ j = 2, \ldots, K+1, \tag{7.6}$$

where $c_j = \frac{N_j}{V}$ is the concentration of component j. Observe that the chemical potential has been replaced by the concentration of the same component in the set.

It is important to note that the set of $(K+2)$ intensive variables is not independent in the sense that there exists an algebraic relationship, the so-called *Gibbs-Duhem relation* among them.

Entropic intensive variables

If, instead of considering the fundamental equation in its energy form (see (7.5)), we consider the entropy S as dependent, we can arrive at the *entropic intensive variables*. The infinitesimal variation form of the so-called entropic fundamental equation (7.2),

$$S = S(U, V, N_1, \ldots, N_K) = S(X_0, X_1, \ldots, X_{K+1}), \tag{7.7}$$

is written in the form:

$$dS = \sum_{j=0}^{K+1} \frac{\partial S}{\partial X_j} dX_j, \tag{7.8}$$

where the canonical set of extensive variables has also been collected in a vector:

$$X_0 = V \ , \ X_1 = U \ , \ X_j = N_{j-1}, \ j = 2, \ldots, K+1. \tag{7.9}$$

Then the entropic intensive variables A_j,

$$A_j = \frac{\partial S}{\partial X_j}, \tag{7.10}$$

are

$$A_0 = \frac{p}{T} \ , \ A_1 = \frac{1}{T} \ , \ A_j = -\frac{\mu_{j-1}}{T}, \ j = 2, \ldots, K+1. \tag{7.11}$$

The entropic intensive variables are also called *thermodynamic driving forces*.

Thermodynamic equations of state

Temperature, pressure and chemical potentials are partial derivatives and are also functions of the canonical set of extensive variables $(S, V, N_1, \ldots, N_K)$:

$$\begin{aligned}
T &= T(S, V, N_1, \ldots, N_K), \\
p &= p(S, V, N_1, \ldots, N_K), \\
\mu_j &= \mu_j(S, V, N_1, \ldots, N_K).
\end{aligned} \tag{7.12}$$

Such relationships are called thermodynamic *equations of state*.

The well-known ideal gas equation that holds for a closed single component region in the form:

$$\frac{p}{T} = R \cdot N \cdot V, \tag{7.13}$$

where R is the universal gas constant, is a simple example of a state equation in the entropy representation.

Euler equation

The fundamental equation (7.4) relates the canonical extensive variables of a system that are all additive when joining regions. Because the internal energy as its dependent variable is also additive, the fundamental equation is a homogeneous first-order equation:

$$U(\lambda S, \lambda V, \lambda N_1, \ldots, \lambda N_K) = \lambda U(S, V, N_1, \ldots, N_K), \tag{7.14}$$

where λ is an arbitrary positive scalar scaling factor. The homogeneous first-order property of the fundamental equation permits it to be written in a particularly convenient form, called the Euler form:

$$U = TS - pV + \mu_1 N_1 + \ldots + \mu_k N_K. \tag{7.15}$$

In entropy representation, the Euler relation takes the form:

$$S = \left(\frac{1}{T}\right) U + \left(\frac{p}{T}\right) V - \sum_{j=1}^{K} \left(\frac{\mu_j}{T}\right) N_j. \tag{7.16}$$

7.1.2 Laws of Thermodynamics

Some of the postulates described in Section 7.1.1 are so important that they deserved the name of "law."

The first (energy conservation) law

Recall that the laws of thermodynamics are postulated for closed systems in equilibrium; therefore, no mass or component mass exchange is considered between the system and its environment. Because thermodynamics was originally concerned with the equilibrium conditions of constrained closed systems when the constraints are removed, it was natural to postulate the general mass and energy conservation laws as the first law of thermodynamics.

For open systems the conservation law of mass, energy and momentum still hold, and one takes them into account in the form of dynamic conservation balances. These constitute the basis of dynamic modelling of process systems, giving rise to the state equations derived from Section 7.2.

The second (entropy) law

Postulate II described in Section 7.1.1 is also called the second (entropy) law of thermodynamics. It has been formulated in various forms over the centuries starting from the impossibility of a perpetuum mobile of the second kind to its rather mathematical form originating from statistical thermodynamics.

Its importance from the viewpoint of passivity-based process control lies in the fact that *there exists an extensive quantity, the entropy, whose increase determines the direction of the changes over the permissible states permitted by the first law.* This approach suggests a search for a relation between the notions of storage function and entropy that is discussed in detail in Section 7.3.

7.1.3 Nonequilibrium Thermodynamics

The classical theory of nonequilibrium thermodynamics (see [43] or [72], for instance) is developed to cover cases when a thermodynamic system evolves in a neighbourhood of its equilibrium point to describe the local processes taking place when moving toward equilibrium. Thus it is an inherently *local* approach that uses linearised relationships. It is out of the scope of this chapter to give even a brief summary of nonequilibrium thermodynamics; only those facets will be briefly recalled that are used to construct a thermodynamically motivated model suitable for passivity-based process control.

Local thermodynamic equilibrium is assumed everywhere in each region. The assumption is standard in the theory of irreversible thermodynamics, and the range of validity covers most of the operating conditions met in practice. It simply states the validity of the entropy form of the Euler equation (7.16) for open systems out of equilibrium.

Onsager relation

The Onsager relation gives the transfer rate $\mathcal{R}_{transfer}$ of conserved extensive quantities as a function of the related thermodynamic driving forces ([72]):

$$\frac{\mathrm{d}X}{\mathrm{d}t} = \mathcal{R}_{transfer} = \mathcal{L}\, q, \tag{7.17}$$

where the matrix $\mathcal{L}$ is *positive definite and symmetrical*, X is a vector of canonical extensive quantities (7.9) and q is the vector of *centred thermodynamic driving force variables*:

$$q_j = A_j - A_j^* \; , \quad j = 0, 1, \ldots, K+1, \tag{7.18}$$

with the entropic intensive variable vector A in (7.11) and the superscript $*$ denoting equilibrium value.

Relationship between thermodynamic variables

In addition to the above, there is a linear static (*i.e.*, time invariant) relationship between the extensive variables and their thermodynamic driving force variables in the form ([52]):

$$q = \mathcal{Q}(X - X^*).\tag{7.19}$$

The coefficient matrix $\mathcal{Q}$ in the above equation is the Hessian of the entropy S with respect to the extensive variables X_j with elements:

$$\mathcal{Q}_{k\ell} = \frac{\partial^2 S}{\partial X_k\, \partial X_\ell} < 0 \quad k, \ell = 0, \dots, K+1.\tag{7.20}$$

The concavity of entropy (see (7.3)) implies that its Hessian with respect to X_j is negative definite; thus $\mathcal{Q}$ is negative definite.

From homogeneity, (7.16) can be written in differential form as follows:

$$dS = \left(\frac{1}{T}\right) dU + \left(\frac{p}{T}\right) dV - \sum_{k=1}^{K} \left(\frac{\mu_k}{T}\right) dN_k,\tag{7.21}$$

where the operator d stands for total derivative. The explicit form of the Hessian (7.20) is obtained by applying the total derivative d to (7.21) resulting in the following matrix:

$$\mathcal{Q} = \begin{bmatrix} -\frac{1}{C_V N T^2} & 0 & 0 & \cdots & 0 \\ 0 & \frac{1}{T}p_V & 0 & \cdots & 0 \\ 0 & 0 & -\frac{\mu_{11}}{T} & \cdots & -\frac{\mu_{1K}}{T} \\ 0 & 0 & \cdots & \cdots & \cdots \\ 0 & 0 & -\frac{\mu_{K1}}{T} & \cdots & -\frac{\mu_{KK}}{T} \end{bmatrix},\tag{7.22}$$

where

$$C_V = \frac{1}{N}\left(\frac{\partial U}{\partial T}\right), \quad p_V = \left(\frac{\partial p}{\partial V}\right), \quad \mu_{jk} = \left(\frac{\partial \mu_j}{\partial c_k}\right)$$

and $N = \sum_{k=1}^{K} N_k$ is the total number of moles.

7.2 The Structure of State Equations of Process Systems

Dynamic system models applied to process control differ substantially from models used in other control fields, such as mechanical, transport, *etc.* Their special nature is explained by the fact that dynamic models of process systems usually originate from dynamic conservation balances for extensive conserved quantities coupled with suitable algebraic constitutive equations of mixed origin. Some of the constitutive equations describe thermodynamic relations such as state equations and thermodynamic property relations.

To explore the implications of the laws of thermodynamics and other process engineering fields, such as reaction kinetics, unit operations *etc.*, for the structure of the state equation, the *standard seven step modelling procedure* [55] will be used that identifies the key steps in process modelling. The main idea behind this procedure is that process models are constructed for a given modelling goal that has implications for the required precision, type of model and mechanisms to be described. The first step in the modelling procedure is to identify the modelling goal clearly together with the boundaries and elements of the process system to be modelled. Thereafter, the mechanisms needed to describe the system are collected in step 2, followed by the review and evaluation of the data available for the model building in step 3. Only then, during step 4, the model equations are constructed, following again a sequence of structured substeps, that starts by identifying all of the balance volumes and then constructing the conservation balances supplemented with algebraic constitutive equations.

If one aims at developing a model for control purposes, a lumped (concentrated parameter) dynamic model is needed, that is of minimal order (*i.e.*, contains a minimum number of state variables) and provides moderate accuracy. Therefore, we restrict our attention to the case of *lumped parameter dynamic process models*.

The material covered here is mainly taken from the book by Hangos and Cameron [55].

7.2.1 State Variables and Order of Systems

The first step in developing a dynamic model of a lumped process system is to identify its balance volumes: They are (quasi-)homogeneous regions that can be regarded perfectly mixed, and they serve as regions over which dynamic conservation balances are constructed. Let **R** be the number of such regions which may be open to the environment and to each other.

As we have seen before in Section 7.1.1, the thermodynamic state of any simple region is unambiguously characterized by the canonical set of extensive variables U, V, $N_1, \ldots, N_K$. Dynamic conservation balances are constructed for the conserved extensive variables that are the internal energy U, the overall mass m and the component masses $m_1, \ldots, m_K$ for a region of lumped parameter process systems (momentum balances occur very rarely, see [55] for a few examples). If one compares the above two sets, it turns out that

- U is present in both sets;
- the component masses m_k can easily be computed from the mole numbers N_k (that are *not* conserved) as $m_k = N_k M_k$ where M_k is the molecular weight;
- the total mass is linearly dependent on the component masses as

$$m = \sum_{j=1}^{K} m_j = \sum_{j=1}^{K} N_j M_j. \tag{7.23}$$

Thus we have $(K+1)$ linearly independent conserved extensive quantities for which balances can be constructed. Their canonical set is usually considered the vector X^e:

$$X_0^e = m, \quad X_1^e = U, \quad X_j^e = m_{j-1}, \quad j = 2, \ldots, K. \quad (7.24)$$

As we shall see later in the next subsection, dynamic conservation balances give rise to the state equations. Therefore, the order of the system is $\mathbf{R}(K+1)$, equal to the number of state variables, if all components are present in each region. The same argument implies that the set of state variables would be the union of the canonical sets of conserved extensive variables of all regions. Because of the difficulties in measuring these variables directly, a related set of engineering intensive variables is used instead that are related to the conserved extensive quantities through *extensive-intensive relationships*, a category of constitutive equations (see Section 7.2.3).

7.2.2 Conservation Balances and Mechanisms

Conservation balances form the differential (prestate) equation part of any process model. These balances are constructed for the conserved extensive variables in each region. The terms in conservation balance equations are related to mechanisms taking place in the system.

The basic equation which drives all the other conservation balances is the *overall mass balance* of a perfectly stirred balance volume:

$$\frac{\mathrm{d}m}{\mathrm{d}t} = v_{in} - v_{out}, \quad (7.25)$$

where v_{in} and v_{out} are the mass inflow and outflow rates, respectively.

A conservation balance equation for another (than overall mass) conserved extensive quantity X_i^e in a perfectly stirred balance volume takes the form

$$\frac{\mathrm{d}X_i^e}{\mathrm{d}t} = v_{in}\chi_{i,in}^e - v_{out}\chi_i^e + \mathcal{R}_{i,transfer} + \mathcal{R}_{i,source}, \quad (7.26)$$

where $\mathcal{R}_{i,transfer}$ is the so-called transfer and $\mathcal{R}_{i,source}$ is the source term. The variable χ_i^e denotes the *mass-specific intensive pair* of the conserved extensive variable X_i^e:

$$\chi_i^e = \frac{X_i^e}{m}, \quad (7.27)$$

where X_i^e can be any element of the vector (7.24) but not the overall mass.

It is important to notice that the general conservation balance equation (7.26) has four terms on its right-hand side corresponding to the four principal mechanisms we take into account when constructing concentrated parameter process models:

- input convection (inflow term, $v_{in}\chi_{i,in}^e$);
- output convection (outflow term, $v_{out}\chi_i^e$);
- (interphase) transfer $\mathcal{R}_{i,transfer}$; and
- sources including both generation and consumption $\mathcal{R}_{i,source}$.

7.2.3 Constitutive Equations

Because of the dependence of the transfer and source terms on intensive variables, one usually needs additional algebraic relationships to complete a process model that has been constructed from dynamic conservation balances. These algebraic relationships are called constitutive equations that are categorized according to their thermodynamic meaning as follows.

1. *Extensive-intensive relationships*
   As we have already seen in Section 7.1.3, an algebraic relationship exists in the form of (7.19) between the thermodynamic extensive variables X_j, $j = 1, \ldots, K+1$ and their intensive thermodynamic potential variables A_j, $j = 1, \ldots, K+1$ that is local and linear in the neighbourhood of the equilibrium point. From this, one can derive approximate linear relationships between the conserved extensive variables X_j^e, $j = 1, \ldots, K+1$ and their intensive engineering counterparts χ_j^e, $j = 1, \ldots, K+1$ by neglecting cross-effects as follows:

$$X_1^e = U = C_P m T, \tag{7.28}$$
$$X_j^e = m_{j-1} = c_{j-1} m \quad , \quad j = 2, \ldots, K+1, \tag{7.29}$$

   where C_P is the specific heat, c_j is the concentration of the jth component, and m is the total mass.

2. *Transfer rate equations*
   Transfer in lumped systems is the transport of conserved extensive quantities through a phase boundary driven by the difference in their intensive (engineering) counterparts. The thermodynamic origin of transfer is the Onsager relationship (see (7.17) in Section 7.1.3) that determines the transfer rate of extensive thermodynamic variables as a linear function of the related thermodynamic driving forces with a positive definite symmetrical matrix $\mathcal{L}$. The Onsager relationship can be applied to derive the engineering version of the transfer rate equations by using local linearisation and neglecting cross-effects. The engineering transfer rate relationship of a conserved extensive quantity X_i^e is in the general linear form:

$$\mathcal{R}_{i,transfer} = K_{transfer} F_{transfer} (\chi_i^{(\ell)} - \chi_i) \tag{7.30}$$

   where $K_{transfer}$ is the transfer rate coefficient (a constant), $F_{transfer}$ is the area of the transfer interphase surface, $\chi_i^{(\ell)}$ is the related engineering driving force variable of the other balance volume and χ_i is that of the actual balance volume over which the conservation balance is constructed.

3. *Source equations*
   The source term in the conservation balance equations is usually highly nonlinear and relates the source of a conserved extensive quantity X_i to the intensive engineering variables (not necessarily only to its related one). The most common source term is the reaction rate expression in the form

$$\mathcal{R}_{i,sourceReact} = V \cdot k_0 \exp\left(\frac{E}{RT}\right) \prod_{k=2}^{K+1} c_k^{\nu_{ik}}, \qquad (7.31)$$

where k_0 is the preexponential factor, E is the activation energy, R is the universal gas constant and ν_{ik} is the stoichiometric coefficient of component k in the ith reaction.

4. *Thermodynamic state equations*

   As we have seen before in Section 7.1.1, the thermodynamic state equation relates some of the intensive variables to the set of canonical extensive variables. In process modelling, state equations occur mainly for balance volumes containing gas phase material.

5. *Physicochemical property relations*

   In the general case, all physicochemical properties, such as densities, heat capacities, reaction enthalpies, transfer rates, *etc.*, are or may be functions of the canonical intensive variables. However, this dependency is most often neglected by *assuming constant physicochemical properties.*

It is important to note that a complete lumped process model contains the conservation balances as differential equations (with their initial values) and the algebraic constitutive equations. Therefore, it is a differential-algebraic (DAE) model. Its differential index [55] characterizes the difficulties related to its solution. It can be shown, however, that under the constant physicochemical property assumption, the resulting process model has index 1, *i.e.*, it falls into the easily solvable category.

7.2.4 State Equations of Process Systems

If one wants to design a controller for a process described by a system model, the above DAE system model should first be transformed to a standard state equation in the form,

$$\frac{\mathrm{d}x}{\mathrm{d}t} = \widetilde{f}(x, u), \qquad (7.32)$$

where x is the state and u is the input variable vector.

The first step in obtaining a state-space model consists of deriving the *intensive form* of the conservation balance equations constructed for the conserved extensive quantities X_i^e to obtain differential equations for the measurable related intensive variable χ_i in the same balance volume with overall mass m. For this purpose, the extensive-intensive relationships in (7.28)–(7.29) are used that are in the form,

$$X_i = K_i m \chi_i, \qquad (7.33)$$

where K_i is a constant. If one differentiates the above equation with respect to time, then,

$$\frac{\mathrm{d}X_i}{\mathrm{d}t} = K_i m \frac{\mathrm{d}\chi_i}{\mathrm{d}t} + K_i \chi_i \frac{\mathrm{d}m}{\mathrm{d}t} \qquad (7.34)$$

and the factor $\frac{dm}{dt}$ in the last term can be substituted in this expression from the mass balance for the same balance volume. Used in this way, the differential variables of the transformed process model are in their intensive form that consists of the following variables for a balance volume:

$$x = [m,\ T,\ c_1,\ \ldots, c_K]^T, \tag{7.35}$$

that form the state vector partition belonging to the balance volume.

Thereafter, one has to *substitute the algebraic equations in the differential ones*, if it is possible. This can always be done if the constant physicochemical property assumption holds and there are no balance equations with gas phase material.

If, in addition, the input variables are chosen in a usual way to have inlet flow rates or intensive variables at the inlets, then the state equation will be in an *input-affine form* as in (2.32):

$$\dot{x} = f(x) + g(x)u. \tag{7.36}$$

Then one can decompose the nonlinear vector-vector functions $f(x)$ and $g(x)$ in the nonlinear state equation in its input-affine form into structurally different additive parts with clear engineering meaning as follows:

$$\dot{x} = \mathcal{A}_{transfer}x + \mathcal{R}_{source}(x) + C_{conv}(u)x + B_{conv}u, \tag{7.37}$$

where $C_{conv}(u)$ is a linear function. The first term in the above equation originates from the transfer, the second from the sources, while the last two correspond to the output and input convection, respectively. The coefficient matrix $\mathcal{A}_{transfer}$ is a constant matrix computed from the matrices $\mathcal{L}$ in (7.17) and $\mathcal{Q}$ in (7.22). The nonlinear source function $\mathcal{R}_{source}(x)$ is of block diagonal form, where the blocks along the state variables belong to the same balance volume.

Thus the decomposed state equation contains a linear state term for the transfer, a general nonlinear state term for the sources, a bilinear or linear input term for the output convection and a linear input convection, respectively.

7.2.5 Implications in Process Control

As a conclusion of the above, one can say that thermodynamics plays a fundamental role in process control by determining the following structural properties of the state equation of lumped dynamic process models:

- *the order of the system* is equal to the number of conserved extensive variables multiplied by the number of balance equations,
- *the set of state variables* that is either the set of conserved extensive variables (extensive form) or the set of their associated engineering intensive pairs,

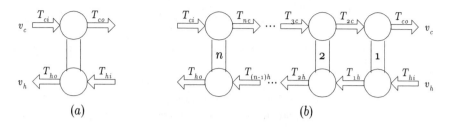

Fig. 7.1. A cascade model of a heat exchanger

- *the number and type of the terms in the state equations* determined by the considered mechanisms, and
- nonlinearities in the state equations.

7.2.6 Heat Exchanger Example

As their name suggests, heat exchangers are used for energy exchange between at least two fluid phase (gas or liquid) streams, a hot and a cold stream. Heat exchangers are usually distributed parameter process systems, but we can build approximate lumped parameter models of them using finite difference approximations of their spatial variables (as in the method of lines approximation scheme [55]). A heat exchanger can then be seen as a composite lumped parameter process system consisting of elementary dynamic units, called *heat exchanger cells*, as depicted in Figure 7.1.

A linear and a nonlinear simple dynamic model of a heat exchanger cell, a single tube-in-shell heat exchanger has already been introduced in Section 2.7. Here we revisit this model and illustrate the principles of physically motivated modelling and passivity-based control using this example.

Engineering model and its variables

From the modelling point of view, a heat exchanger cell consists of two perfectly stirred (lumped) balance volumes (called *lumps*) connected by a heat conducting wall. One of the lumps is called the hot (h) side and the other one is the cold (c) side. The lumps with their variables are shown in Figure 7.1a (or before in Figure 2.8 of Section 2.7).

We assume constant overall mass, constant physicochemical properties and a single component in both balance volumes. The constant overall mass assumption implies that both the inlet and outlet mass flow rates are equal for each balance volume.

Therefore, only dynamic *energy conservation balances* should be constructed for this system with input and output convection and a transfer term but no source term in the following form:

$$\frac{dU_c(t)}{dt} = v_c(t)c_{Pc}\left[T_{ci}(t) - T_c(t)\right] + K_T F\left[T_h(t) - T_c(t)\right], \qquad (7.38)$$

$$\frac{dU_h(t)}{dt} = v_h(t)c_{Ph}\left[T_{hi}(t) - T_h(t)\right] + K_T F\left[T_c(t) - T_h(t)\right], \qquad (7.39)$$

where $v_\bullet$ is the mass flow rate, $T_{\bullet i}$ is the inlet temperature, F is the heat transfer area and K_T is the heat transfer coefficient of the lump $\bullet \in \{h, c, \}$.

The *constitutive* equations include the engineering energy transfer term:

$$\mathcal{R}_{transfer} = K_T F\left[T_h(t) - T_c(t)\right], \qquad (7.40)$$

that has already been substituted into (7.38)–(7.39). The engineering form of extensive-intensive relationship,

$$U_\bullet = c_{P\bullet}m_\bullet T_\bullet, \qquad (7.41)$$

with $\bullet \in \{h, c, \}$ accompanies the equations that can be used to develop the intensive form of the balance equations.

The *thermodynamic model variables* are as follows. The vectors of conserved extensive variables and their related intensive pairs are

$$X^e = [\, U_c, \ U_h \,]^T, \quad x = [\, T_c, \ T_h \,]^T, \qquad (7.42)$$

while the thermodynamic driving force vector is

$$A^e = \left[\, \frac{1}{T_c}, \ \frac{1}{T_h} \,\right]^T. \qquad (7.43)$$

Thus the negative definite coefficient matrix in the extensive-intensive relationship,

$$A^e - A^{e*} = q = \mathcal{Q}\left(X^e - X^{e*}\right), \qquad (7.44)$$

is in the form,

$$\mathcal{Q} = \begin{bmatrix} -\dfrac{1}{c_{Pc}m_c(T^*)^2} & 0 \\ 0 & -\dfrac{1}{c_{Ph}m_h(T^*)^2} \end{bmatrix}, \qquad (7.45)$$

with an equilibrium reference temperature T^*. The above equation implies a relationship between the temperatures and the thermodynamic driving force variables together with (7.41):

$$T_\bullet - T^* = -T^{*2}\left(\frac{1}{T_\bullet} - \frac{1}{T^*}\right), \quad \bullet \in \{h, c, \}. \qquad (7.46)$$

State-space model and system variables

With our modelling assumptions, the intensive form of of the conservation balances is obtained by substituting the extensive-intensive relationship (7.41) in the energy conservation balances:

$$\frac{dT_c}{dt} = \frac{v_c}{m_c}(T_{ci} - T_c) + \frac{K_T}{c_{Pc}m_c}F(T_h - T_c), \tag{7.47}$$

$$\frac{dT_h}{dt} = \frac{v_h}{m_h}(T_{hi} - T_h) + \frac{K_T}{c_{Ph}m_h}F(T_c - T_h). \tag{7.48}$$

The above equations serve as the state equations where the state and possible input or disturbance variable vectors are as follows:

$$x = [\ T_c,\ T_h\ ]^T, \quad u = [\ v_c,\ v_h,\ T_{ci},\ T_{hi}\ ]^T. \tag{7.49}$$

7.3 Physically Motivated Supply Rates and Storage Functions

To focus on the thermodynamic background of passivity-based process control, we restrict ourselves to a simple class of lumped parameter (finite dimensional) process systems that obey the following *basic modelling assumptions*:

1. perfectly stirred (lumped) balance volumes,
2. constant physicochemical properties, and
3. constant pressure.

Note that the last two assumptions are related and they ensure a simple form of the convection term in the conservation balance equations. With the second and third assumptions, the algebraic constitutive equations can always be substituted in the conservation balances.

7.3.1 Entropy-based Storage Functions

First the notion of the storage function of passive systems introduced in Section 2.1 is repeated here for convenience. A scalar function $\mathcal{S}$ of the state variables x is called a storage function if

$$\mathcal{S}(x(t)) \leq \mathcal{S}(x(0)) + \int_0^t u^T(s)y(s)ds, \tag{7.50}$$

with $\mathcal{S}(x) \geq 0$ if $x \neq 0$ and $\mathcal{S}(x) = 0$ if $x = 0$. Here the number of system inputs should be equal to the number of system outputs, *i.e.*, $\dim u = \dim y$, and the scalar time-dependent quantity $w(u, y) = u^T y$ is called the *supply rate*.

Now let us briefly recall the most important elements of the thermodynamic description of process systems relevant to the construction of an entropy-based storage function.

1. The state of a balance volume is characterized by the canonical set of conserved extensive variables $X^e = [U, (m_j, j = 1, \dots, K)]$, with $m = \sum_{j=1}^{K} m_j$.

2. There exists a concave entropy function $S(X^e)$.
3. $S(X^e)$ is homogeneous of degree one, *i.e.*, $S(\lambda X^e) = \lambda S(X^e)$.
4. There exists a conserved extensive quantity X_i^e (usually the internal energy U) for which $\frac{\partial S}{\partial X_i^e} > 0$.

We have also associated entropic intensive variables or driving forces A_i^e with each of the conserved extensive quantities through the defining equation:

$$A_i^e = \frac{\partial S}{\partial X_i^e}, \quad A^e = \left[\frac{1}{T}, \left(-\frac{\mu_j}{T}, j = 1, \dots, K\right)\right]. \tag{7.51}$$

Defining equations

With the above properties the scalar function,

$$\mathcal{S}^e(X^e) = -S + S^* + A^{*T}(X^e - X^{e*}) = -(A^e - A^{e*})^T X^e = -q^T X^e, \tag{7.52}$$

is a natural storage function candidate, where S^* is the entropy in an equilibrium state and q is the centred version of the entropic intensive variables A^e above. The last equality holds as a consequence of the linear static relationship (7.19) between the conserved extensive variables and their related intensive potential variables.

One can *generalize* the above storage function construction as follows [64]: First we generalize the storage function with a state dependent term,

$$\mathsf{s}(x(t)) \leq \mathsf{s}(x(0)) + \int_0^t u^T(s)y(s)ds - \epsilon_0 \int_0^t x^T(s)x(s)ds, \tag{7.53}$$

where $\epsilon_0 > 0$ is a scalar constant and $\mathsf{w}(u, y, x) = u^T y - \epsilon_0 x^T x$ is the generalized supply rate. Note that the generalized entropy-based storage function s can be obtained from the original one by adding a positive definite term to it, *i.e.*,

$$\mathsf{s}(x(t)) = \mathcal{S}(x(t)) + \epsilon_0 \int_0^t x^T(s)x(s)ds. \tag{7.54}$$

This construction helps to prove the passivity of process systems with more than one connected balance volume, *i.e.*, of *process networks* [64].

Consider a state evolution (a solution of the system model) (X^{e1}, A^{e1}) and another evolution (X^{e2}, A^{e2}). Then the scalar function,

$$\mathsf{s}^e(X^{e1}, X^{e2}) = (X^{e1} - X^{e2})^T (A^{e2}(X^{e2}) - A^{e1}(X^{e1})), \tag{7.55}$$

is a generalized version of the function $\mathcal{S}^e$ of type (7.53), where $X^{e1} = X^{*e}$ is an equilibrium constant reference. Defining deviation variables,

$$\overline{X^e} = \begin{bmatrix} X^{e1} \\ X^{e2} \end{bmatrix}, \quad \overline{A^e} = \begin{bmatrix} A^{e1} - A^{e2} \\ A^{e2} - A^{e1} \end{bmatrix}, \tag{7.56}$$

(7.55) can be rewritten as

$$\mathbf{s}^e(\overline{X^e}) = -\overline{A^e}^T \overline{X^e}. \tag{7.57}$$

In the case of $\mathbf{R}$ balance volumes, one can simply sum the storage functions corresponding to the individual balance volumes, such that

$$\mathcal{S}^e = \sum_{j=1}^{\mathbf{R}} \mathcal{S}^{e(j)}, \tag{7.58}$$

where $\mathcal{S}^{e(j)}$ is the storage function of the jth balance volume.

Properties

First we check if both the simple entropy-based storage function candidate $\mathcal{S}^e$ in (7.52) and its generalized version $\mathbf{s}^e$ in (7.57) possesses the required positivity properties, i.e., $\mathcal{S}(\xi) > 0$ if $|\xi| \neq 0$ and $\mathcal{S}(\xi) = 0$ if $|\xi| = 0$.

The definition given in (7.52) implies the required properties if the independent variable is chosen as the centred engineering conserved variables, i.e., $\xi = X^e - X^{e*}$.

- At the equilibrium point $\xi = 0$,

$$\mathcal{S}^e(0) = \mathcal{S}^e(X^{e*}) = -S^* + S^* + A^{*T}(X^{e*} - X^{e*}) = 0. \tag{7.59}$$

- Outside equilibrium, when $|\xi| \neq 0$, one can approximate the entropy with its Taylor series expansion around the equilibrium value:

$$S(\xi) = S^* + A^{*T}\xi + \text{(higher order terms)}, \tag{7.60}$$

where the higher order terms are negative because of the concavity of entropy. This implies that the storage function candidate $\mathcal{S}^e = S^* + A^{*T}\xi$ is positive when the system is not at equilibrium.

Thus the properties of entropy guarantee that the storage function candidates are convex functions of their independent variables and take their minimal value of zero at the equilibrium state or at the reference solution.

The entropy-like definition of the storage function candidate $\mathcal{S}^e$ enables us to connect the state-space model equations to the defining property of the storage function in the form,

$$\frac{d\mathcal{S}^e}{dt} \leq u^T y, \tag{7.61}$$

that will help us to identify system inputs and outputs for which the process system is passive. If one differentiates (7.52) with respect to time, the following form is obtained:

$$\frac{\mathrm{d}\mathcal{S}^e}{\mathrm{d}t} = -q^T \frac{\mathrm{d}X^e}{\mathrm{d}t}, \tag{7.62}$$

taking into account the homogeneity of the entropy function. The time derivative $\frac{\mathrm{d}X^e}{\mathrm{d}t}$ connects the time derivative of the storage function to the conservation balance equations or state equations in their extensive form thus enabling the substitution of the right-hand side of the conservation balances containing the input variables in (7.62). Therefore, the check of the defining property (7.61) can be easily performed by using (7.62).

7.3.2 Possible Choices of Inputs and Outputs

If one wants to select suitable inputs and outputs for a given process system such that it becomes passive with respect to the entropy-based storage function candidate $\mathcal{S}^e$, then the following constraints should be respected:

1. $\dim u = \dim y$ to have a scalar supply rate $w(u, y) = u^T y$, and
2. the unit of the supply rate w should be $\frac{J}{sec \cdot K}$.

More insight can be gained if one compares the right-hand sides of the inequality (7.61) and the equality (7.62) for $\frac{\mathrm{d}\mathcal{S}^e}{\mathrm{d}t}$. It suggests the use of a scaled version of centred driving force variables $q = A^e - A^{*e}$ as output, and the scaled version of the inlet flow of the corresponding conserved extensive quantity $v_{in}X_{in}^e$ (see (7.26)) or its centred version as its pair; then $\dim x = \dim u$. In the simplest case, it means that the intensive engineering variables at the inlet of the system χ_{in}^e are selected as input variables and the inlet mass flow rates are kept constant. Then the outputs are assumed to be $y = A^e$, where A^e is the set of engineering driving force variables that can be re-scaled to have either the set of conserved extensive quantities or their related engineering intensive variables.

7.3.3 Storage Function of the Heat Exchanger Example

We have already constructed the engineering model of the simple heat exchanger cell example in Section 7.2.6, that will be augmented here with an entropy-based storage function.

Entropy-based storage function

A joint equilibrium reference temperature T^* for both balance volumes is used for the construction to have a special case of the entropy-based storage function $\mathcal{S}^e$ in (7.52):

$$\mathcal{S}^e(U_c, U_h) = \sum_{j=c,h} -S_j + S_j^* + \frac{1}{T^*}(U_j - U_j^*) = -\sum_{j=c,h}\left(\frac{1}{T_j} - \frac{1}{T^*}\right)U_j. \tag{7.63}$$

Its differential form is

$$\frac{\mathrm{d}\mathcal{S}^e}{\mathrm{d}t} = -\sum_{j=c,h} \left(\frac{1}{T_j} - \frac{1}{T^*} \right) \frac{\mathrm{d}U_j}{\mathrm{d}t}, \tag{7.64}$$

where one can directly substitute the right-hand side of the conservation balances (7.38)–(7.39).

Input and output variables and passivity

Following the recommendations in Section 7.3.2, the hot and cold inlet temperatures T_{ci} and T_{hi} are selected as physical input variables giving rise to the following input and output vectors:

$$u = [\, \kappa_c \cdot T_{ci}, \;\; \kappa_h \cdot T_{hi} \,]^T, \quad y = \left[\frac{1}{T_c} - \frac{1}{T^*}, \;\; \frac{1}{T_h} - \frac{1}{T^*} \right]^T, \tag{7.65}$$

where $\kappa_c = c_{Pc} v_c$ and $\kappa_h = c_{Ph} v_h$ are constants.

With the above choice, one can show the passivity of the heat exchanger cell with respect to the input and output variables in (7.65) using the differential form of the entropy-based storage function (7.64). First, we form vectors from the centred state variables and their centred driving force variables by observing that

$$\overline{T} = \begin{bmatrix} T_c - T^* \\ T_h - T^* \end{bmatrix}, \tag{7.66}$$

$$y = q = \begin{bmatrix} \frac{1}{T_c} - \frac{1}{T^*} \\ \frac{1}{T_h} - \frac{1}{T^*} \end{bmatrix} = -\frac{1}{T^{*2}} \begin{bmatrix} T_c - T^* \\ T_h - T^* \end{bmatrix} = -\frac{1}{T^{*2}} \overline{T}. \tag{7.67}$$

Note that the above relationship between the centred driving force variable y_c (or y_h) and its corresponding centred state variable $\overline{T}_c$ (or $\overline{T}_h$) is obtained by expanding the function $f(T) = 1/T$ into a Taylor series around the equilibrium point T^* and considering a linear approximation in the form,

$$f(T) \cong \frac{1}{T^*} - \frac{1}{T^{*2}}(T - T^*). \tag{7.68}$$

By substituting the right-hand side of the conservation balances (7.38)–(7.39) in the differential form, the following equation is obtained:

$$\begin{aligned}
\frac{\mathrm{d}\mathcal{S}^e}{\mathrm{d}t} &= \frac{1}{T^{*2}} \overline{T}^T \frac{\mathrm{d}X^e}{\mathrm{d}t} \\
&= \frac{1}{T^{*2}} \overline{T}^T \begin{bmatrix} -K_T F & K_T F \\ K_T F & -K_T F \end{bmatrix} \overline{T} + \frac{1}{T^{*2}} \overline{T}^T \begin{bmatrix} -v_c c_{Pc} & 0 \\ 0 & -v_h c_{Ph} \end{bmatrix} \overline{T} + y^T u.
\end{aligned} \tag{7.69}$$

The first and second terms on the right-hand side of the above equation originate from the transfer and output convection terms, respectively, and they are both nonpositive and negative, respectively, being quadratic forms with negative (semi)definite matrices. This implies that

$$\frac{\mathrm{d}\mathcal{S}^e}{\mathrm{d}t} < y^T u \tag{7.70}$$

in this case, that is, *the simple heat exchanger cell is passive with respect to the input and output variables (7.65).*

7.4 Hamiltonian Process Models

The notion of Hamiltonian systems (see *e.g.*, [130]) has been abstracted from the principles of theoretical mechanics where it has become clear that the underlying physics determines a special nonlinear structure that can be used effectively in nonlinear systems and control theory. The notions and notations used in this section are based on the book by van der Schaft [130].

7.4.1 System Structure and Variables

The so-called Hamiltonian control systems have the form,

$$\dot{q} = \frac{\partial H^T}{\partial p}(q, p), \tag{7.71}$$

$$\dot{p} = -\frac{\partial H^T}{\partial q}(q, p) + B(q)u, \tag{7.72}$$

$$y = B^T(q)\frac{\partial H^T}{\partial p}(q, p), \tag{7.73}$$

where $q \in \mathbb{R}^k$ are the generalized configuration coordinates, $p \in \mathbb{R}^k$ are the generalized momenta, $u \in \mathbb{R}^m$ is the input, $B(q) \in \mathbb{R}^{k \times m}$ is the input force matrix and $H : \mathbb{R}^{2k} \mapsto \mathbb{R}$ is the Hamiltonian function. If $m < k$, then the system (7.71)–(7.73) is called underactuated, while with $k = m$ and $B(q)$ invertible everywhere, it is called fully actuated. The coordinates q and p are often called states and costates of the system, respectively.

In classical mechanics, $B(q)u$ is the vector of generalized forces acting on the system, while H, the total energy of the system, is given by

$$H(q, p) = \frac{1}{2}p^T M^{-1}(q)p + V(q), \tag{7.74}$$

where $M(q)$ is a $k \times k$ inertia (or generalized mass) matrix which is symmetrical and positive definite. The first term on the right-hand side of (7.74) is the kinetic energy while the second term V denotes the potential energy. Using the relation $p = M(q)\dot{q}$, the Hamiltonian function can also be written as

$$H(q,p) = \frac{1}{2}\dot{q}^T M(q)\dot{q} + V(q).\tag{7.75}$$

Taking the time derivative of the Hamiltonian function gives:

$$\begin{aligned}
\dot{H} &= \frac{\partial H}{\partial q}(q,p)\dot{q} + \frac{\partial H}{\partial p}(q,p)\dot{p} \\
&= \frac{\partial H}{\partial p}(q,p)\left(-\frac{\partial H^T}{\partial q}(q,p) + B(q)u\right) + \frac{\partial H}{\partial q}(q,p)\frac{\partial H^T}{\partial p}(q,p) \\
&= \frac{\partial H}{\partial p}(q,p)B(q)u = y^T u.
\end{aligned}\tag{7.76}$$

It can be seen from (7.76) that H is a conserved quantity (first integral) for $u = 0$. For arbitrary u, the system is lossless (see Definition 2.8) with the output defined in (7.73) and storage function H. If the geometry of H is appropriate, then it can be used as a Lyapunov function for the open-loop system or as a control Lyapunov function when designing a feedback control system that makes H strictly decreasing.

7.4.2 Generalized Hamiltonian Systems

Observe that the number of state variables of Hamiltonian control systems is always even. This can make finding the Hamiltonian representation of non-mechanical systems difficult. A generalized version of (7.71)–(7.73) that preserves the most important control related dynamical properties can be given in the form,

$$\dot{x} = J(x)\frac{\partial H^T}{\partial x}(x) + g(x)u,\tag{7.77}$$

$$y = g^T(x)\frac{\partial H^T}{\partial x}(x),\tag{7.78}$$

where $x \in \mathbb{R}^n$ is the state vector, $H : \mathbb{R}^n \mapsto \mathbb{R}$ is the Hamiltonian function, $u, y \in \mathbb{R}^m$ are the input and output vectors, respectively, $g(x) \in \mathbb{R}^{n \times m}$, and $J(x)$ is an $n \times n$ skew symmetrical (i.e., $J(x) = -J(x)^T$) matrix smoothly depending on x.

The time derivative of H now reads:

$$\begin{aligned}
\dot{H} &= \frac{\partial H}{\partial x}(x) \cdot \dot{x} \\
&= \frac{\partial H}{\partial x}(x)J(x)\frac{\partial H^T}{\partial x}(x) + \frac{\partial H}{\partial x}(x)g(x)u.
\end{aligned}\tag{7.79}$$

From the skew-symmetry of J, it follows that

$$\dot{H} = \frac{\partial H}{\partial x}(x)g(x)u = y^T u,\tag{7.80}$$

which means that (7.77)–(7.78) are also lossless with storage function H.

7.4.3 Generalized Hamiltonian Systems with Dissipation

The above defined Hamiltonian and generalized Hamiltonian systems are both lossless (which is a rather small system class from a practical point of view), while it is known that the majority of real process systems are dissipative in nature. To transform the generalized Hamiltonian systems into possibly dissipative systems, let us define the following linear partial feedback:

$$u_R = -Ky_R, \tag{7.81}$$

where K is a positive semidefinite symmetrical matrix and

$$u = \begin{bmatrix} u_0 \\ u_R \end{bmatrix}, \quad g(x) = \begin{bmatrix} g_0(x) \ g_R(x) \end{bmatrix}, \tag{7.82}$$

$$y = \begin{bmatrix} y_0 \\ y_R \end{bmatrix} = \begin{bmatrix} g_0^T(x)\frac{\partial H^T}{\partial x}(x) \\ g_R^T(x)\frac{\partial H^T}{\partial x}(x) \end{bmatrix}. \tag{7.83}$$

The closed-loop system is described by the equations

$$\dot{x} = [J(x) - R(x)]\frac{\partial H^T}{\partial x}(x) + g_0(x)u_0, \tag{7.84}$$

$$y_0 = g_0^T(x)\frac{\partial H^T}{\partial x}(x), \tag{7.85}$$

where

$$R(x) = g_R(x)Kg_R^T(x) \tag{7.86}$$

is a positive semidefinite symmetrical matrix, called the *dissipation matrix*. It is important to note that any real square matrix W can be decomposed as a sum of a skew-symmetrical and a symmetrical matrix:

$$W = \frac{1}{2}\left(W - W^T\right) + \frac{1}{2}\left(W + W^T\right). \tag{7.87}$$

Taking the time derivative of H along the solutions of (7.84) and (7.85) gives

$$\dot{H} = y_0^T u_0 - \frac{\partial H^T}{\partial x}(x)R(x)\frac{\partial H}{\partial x}(x) \leq y_0^T u_0, \tag{7.88}$$

which shows that the system represented by (7.84) and (7.85) is passive.

Note that for system equations written in original physical coordinates, $J(x)$ often reflects the system's energy preserving internal interconnection structure, while $R(x)$ is related to an additional resistive structure. The generalized Hamiltonian structure (with dissipation) has been identified in many kinds of physical systems such as electric circuits [82], electromechanical systems [90], *etc.*

7.4.4 Hamiltonian Description of the Heat Exchanger Example

Linear model

Consider the linear heat exchanger model and the storage function S described in Section 2.7. Let us use S as the Hamiltonian function, *i.e.*, $H = S$. The gradient of H is written as

$$\frac{\partial H}{\partial x}(x) = \begin{bmatrix} \frac{1}{k_1}x_1 & \frac{1}{k_2}x_2 \end{bmatrix}. \tag{7.89}$$

Let us introduce the following notation:

$$M(x) = J(x) - R(x). \tag{7.90}$$

It is easy to see that the generalized Hamiltonian form of (2.121) can be constructed as

$$\dot{x} = M(x) \cdot \frac{\partial H^T}{\partial x}(x) + \begin{bmatrix} a_1 & 0 \\ 0 & a_2 \end{bmatrix} u, \tag{7.91}$$

where

$$M(x) = \begin{bmatrix} k_1(-a_1 - k_1) & k_1 k_2 \\ k_1 k_2 & k_2(-a_2 - k_2) \end{bmatrix}, \tag{7.92}$$

i.e., in this case, M is a symmetrical constant matrix. To examine the derivative of H, we have to check the definiteness of M. The first minor of M is clearly negative: $M_{11} = k_1(-a_1 - k_1)$ (recall that the parameters a_i and k_i are positive). The second minor is the determinant,

$$\det(M) = a_1 a_2 k_1 k_2 + a_1 k_1 k_2^2 + a_2 k_1^2 k_2, \tag{7.93}$$

which is positive. From this, it follows that M is negative definite, so H is always decreasing for $u = 0$. Since M is symmetrical, the matrices J and R are

$$J = 0^{2 \times 2}, \quad R = -M. \tag{7.94}$$

The passive output vector of the system can be calculated as

$$y = \begin{bmatrix} a_1 & 0 \\ 0 & a_2 \end{bmatrix} \cdot \begin{bmatrix} \frac{1}{k_1}x_1 \\ \frac{1}{k_2}x_2 \end{bmatrix} = \begin{bmatrix} \frac{a_1}{k_1}x_1 \\ \frac{a_2}{k_2}x_2 \end{bmatrix}. \tag{7.95}$$

This means that the linear heat exchanger model admits a generalized dissipative Hamiltonian description that exists independently of the parameter values.

Nonlinear model

Now, let us consider the nonlinear heat exchanger model described in (2.132). Let us introduce the following notations:

$$b_1 = \frac{1}{V_c} u_{10}, \quad b_2 = \frac{1}{V_h} u_{20}. \tag{7.96}$$

We use the storage function $S(x')$ again as the Hamiltonian function *i.e.*,

$$H(x') = \frac{1}{2} \left(\frac{1}{k_1} x_1'^2 + \frac{1}{k_2} x_2'^2 \right). \tag{7.97}$$

The generalized Hamiltonian description can be written as

$$\dot{x}' = M(x') \cdot \frac{\partial H^T}{\partial x'}(x') + G(x')u', \tag{7.98}$$

where

$$M(x') = \begin{bmatrix} k_1(-b_1 - k_1) & k_1 k_2 \\ k_1 k_2 & k_2(-b_2 - k_2) \end{bmatrix} \tag{7.99}$$

and

$$G(x') = \begin{bmatrix} \frac{1}{V_c}(T_{ci} - x_1' - x_{10}) & 0 \\ 0 & \frac{1}{V_h}(T_{hi} - x_2' - x_{20}) \end{bmatrix}. \tag{7.100}$$

By comparing (7.92) and (7.99), it is clear that the structure and sign conditions in $M(x)$ and $M(x')$ are the same. Therefore, $M(x')$ is also uniformly negative definite.

However, the passive output in this case is different from the linear case because of the difference in the input structure of the two models. The output can now be calculated as

$$\begin{aligned}
y' &= \begin{bmatrix} \frac{1}{V_c}(T_{ci} - x_1' - x_{10}) & 0 \\ 0 & \frac{1}{V_h}(T_{hi} - x_2' - x_{20}) \end{bmatrix} \cdot \begin{bmatrix} \frac{1}{k_1} x_1' \\ \frac{1}{k_2} x_2' \end{bmatrix} \\
&= \begin{bmatrix} \frac{1}{k_1 V_c}(T_{ci} x_1' - x_1'^2 - x_{10} x_1') \\ \frac{1}{k_2 V_h}(T_{hi} x_2' - x_2'^2 - x_{20} x_2') \end{bmatrix}.
\end{aligned} \tag{7.101}$$

This completes the generalized Hamiltonian description of the nonlinear heat exchanger model.

7.5 Case Study: Reaction Kinetic Systems

The (possibly complex) dynamics of a biochemical reaction network has been a target of intensive research for decades. However, there are well-established stability results for closed reaction kinetic systems that can be used for establishing useful passivity results. This section is based mainly on [92].

7.5.1 System Description, Thermodynamic Variables and State-space Model

To describe the basic structure of reaction kinetic systems, we use the notation employed in [45].

The isolated and homogeneous isothermal systems where n chemical species participate in an r-step reaction network, are represented by the following stoichiometric mechanism:

$$\sum_{i=1}^{n} \alpha_{ij} A_i \rightleftarrows \sum_{i=1}^{n} \beta_{ij} A_i \quad j = 1, \ldots, r, \tag{7.102}$$

where α_{ij}, β_{ij} are the constant stoichiometric coefficients for species A_i in the reaction step j. All reactions are assumed to be reversible, and reaction rates obey the *mass action law* [39]:

$$W_j = W_j^+ - W_j^- = k_j^+ \prod_{i=1}^{n} x_i^{\alpha_{ij}} - k_j^- \prod_{i=1}^{n} x_i^{\beta_{ij}}, \tag{7.103}$$

where k_j^+ and k_j^- are the constants of the direct and inverse rates of the jth reaction step, respectively, and $x_i \geq 0$ represents the concentration of species A_i. Each concentration evolves in time according to the ordinary differential equation,

$$\dot{x}_i = \sum_{j=1}^{r} \nu_{ij} (W_j^+ - W_j^-), \tag{7.104}$$

where $\nu_{ij} = \alpha_{ij} - \beta_{ij}$ is positive or negative depending on whether the specie i is a product or a reactant in the reaction j. The dynamic evolution of the network can then be represented by a set of ordinary differential equations which in compact matrix form is written as

$$\dot{x} = \mathcal{N} \cdot W(x), \tag{7.105}$$

where $\mathcal{N} = [\nu_{ij}]$ is the $n \times r$ coefficient matrix whose columns are the linearly independent stoichiometric vectors $\nu_j = \beta_j - \alpha_j$, and $W(x) \in \mathbb{R}^r$ denotes the vector of reaction rates.

7.5.2 The Reaction Simplex and the Structure of Equilibrium Points

It is easy to see from the structure of (7.105) that the linearly independent functions,

$$c_k^T \cdot x, \quad k = 1, \ldots, m, \tag{7.106}$$

are invariant for the dynamics of the system, where $c_k \in \mathbb{R}^n$, $k = 1, \ldots, m$ form a basis of $\ker(\mathcal{N}^T)$. These invariant quantities reflect the conservation laws of

closed kinetic networks, where the amount of each radical or chemically un-
altered component remains constant and the corresponding component mass
balances satisfy $c_k^T \cdot x(t) = c_k^T \cdot x(0)$.

It is known that the positive orthant $\mathbb{R}_+^n$ of the concentration space is
also invariant for (7.105) since the concentrations are always nonnegative.
Therefore, we can define the so-called reaction simplex (which is also invariant
for the system's dynamics) corresponding to an initial concentration vector
x_0 as the intersection of the conservation laws and the positive orthant:

$$\Omega(x_0) = \left\{ x \in \mathbb{R}_+^n \mid c_k^T \cdot x = x_0 \right\}. \tag{7.107}$$

It is shown (see [39] or [117]) that in a closed reaction network, there is a
unique equilibrium point of (7.105) on each reaction simplex which is stable
in the space of concentrations (and asymptotically stable if we restrict the
dynamics to the given reaction simplex).

7.5.3 Physically Motivated Storage Function

The second law of thermodynamics establishes an evolution criterion based on
a concave function (the entropy) which never decreases in isolated systems and
achieves its maximum at equilibrium (see the second law of thermodynamics
in Section 7.1.2). Isolated dissipative systems evolve to equilibrium through
irreversible processes that produce entropy. The rate of entropy production
is a way to quantify dissipation. In closed reaction systems, the total mass,
volume (V) and energy (U) are constant. Therefore, the total entropy $\tilde{S}$ of
the system can be written in the form of the Euler equation (7.16) [25]:

$$\tilde{S} = \left(\frac{1}{T}\right) U + \left(\frac{P}{T}\right) V - \sum_{i=1}^n \left(\frac{\tilde{\mu}_i}{T}\right) M_i, \tag{7.108}$$

where T is the temperature, P is the pressure and $\tilde{\mu}_i$ is the chemical potential
of the ith component with component mass M_i. Assuming in addition constant temperature and ideal mixtures, where the chemical potential of the ith
component is $\tilde{\mu}_i = RT \ln x_i$ and its mass is $M_i = V x_i$ we obtain

$$\tilde{S} = \left(\frac{1}{T}\right) U + \left(\frac{P}{T}\right) V - \frac{RV}{T} \sum_{i=1}^n (\ln x_i) x_i, \tag{7.109}$$

where R is the universal gas constant.

In exploring the stability of the equilibrium manifold, we follow [45] and
define an entropy-like expression which coincides with the negative of the
function,

$$\mathcal{V}(x) = \sum_{i=1}^n x_i \cdot (\ln x_i - 1) = \sum_{i=1}^n (x_i \ln x_i - x_i), \tag{7.110}$$

where the first term in the summand corresponds to the last term and the second to the second one in (7.109). In addition, the right-hand side term in (7.105) is Lipschitz continuous. This implies, as discussed in [1], that for any arbitrary reference x_1 there exists a nonnegative function $L_\lambda(x, x_1)$ associated with a constant $\lambda \geq 0$, such that the following relation holds:

$$[\mu(x) - \mu(x_1)]^T \mathcal{N} \cdot W(x) + L_\lambda(x, x_1)$$
$$= \lambda[\mu(x) - \mu(x_1)]^T (x - x_1), \qquad (7.111)$$

where $\mu = \nabla_x \mathcal{V} = \ln x$ and the logarithm is taken elementwise. Systems which, in addition, satisfy $L_0(x, x_1) \geq 0$, are *purely dissipative*. Closed reaction networks are a class of purely dissipative systems for a state x_1 that is the equilibrium reference. This can be easily shown by noting that for $\lambda = 0$ and $x_1 = x^*$ in (7.111),

$$L_0(x, x^*) = -(\mu - \mu^*)^T \sum_{j=1}^r \nu_j W_j. \qquad (7.112)$$

Equation 7.112 can be written in terms of direct and inverse reaction rates (7.103) as

$$L_0(x, x^*) = \sum_{j=1}^r \ln \frac{W_j^+}{W_j^-} \cdot (W_j^+ - W_j^-). \qquad (7.113)$$

Since each term on the right-hand side of (7.113) is nonnegative, we conclude that $L_0(x, x^*) \geq 0$.

To derive the stability conditions for closed reaction networks, we define a positive definite and convex function $S(x)$, constructed as the difference between $\mathcal{V}(x)$ and its supporting hyperplane at the equilibrium reference x^*:

$$S(x) = \sum_{i=1}^n x_i \left(\ln \frac{x_i}{x_i^*} - 1 \right) + x_i^*. \qquad (7.114)$$

Taking the time derivative of S along the solutions of (7.105) and using (7.111) with $\lambda = 0$, we obtain

$$\dot{S} = (\mu - \mu^*)^T \sum_{j=1}^r \nu_j W_j = -L_0(x, x^*). \qquad (7.115)$$

Since $L_0 \geq 0$ by (7.113), $\dot{S} \leq 0$. Consequently, S is a legitimate Lyapunov function that ensures the structural stability of the reaction network at the equilibrium reference.

7.5.4 Passive Input-output Structure

The state-space representation of an open reaction system, *i.e.*, a system which exchanges mass with its environment, is constructed by adding a set of input and output convection terms (see Section 7.2.2) to the closed reaction system (7.105). To handle standard operating conditions in chemostats, we further assume that the overall mass of the system is kept constant by having the same input and output overall mass flow Φ. In this way, the set of ordinary differential equations governing the evolution of states becomes

$$\dot{x} = \mathcal{N} \cdot W(x) + \phi(x^{in} - x), \tag{7.116}$$

where $\phi = \Phi/V$ (with V as the volume of the chemostat) denotes the inverse of the residence time and x^{in} is the inlet concentration vector.

If the input is the inverse residence time, $i.e.$, $u = \phi$ (a single input which is proportional to the input/output mass flow rate), then the components of the input affine state-space model $\dot{x} = f(x) + g(x)u$ are

$$f(x) = \mathcal{N} \cdot W(x), \quad g(x) = x^{in} - x. \tag{7.117}$$

Now, the output can be chosen so that the system satisfies the Kalman–Yacubovitch–Popov property:

$$
\begin{aligned}
y = h(x) = L_g S(x) &= \frac{\partial S}{\partial x} g(x) \\
&= \begin{bmatrix} \dfrac{\partial S}{\partial x_1} & \cdots & \dfrac{\partial S}{\partial x_n} \end{bmatrix} \begin{bmatrix} x_1^{in} - x_1 \\ \vdots \\ x_n^{in} - x_n \end{bmatrix},
\end{aligned} \tag{7.118}
$$

where

$$\frac{\partial S}{\partial x_i} = \ln\left(\frac{x_i}{x_i^*}\right), \quad x_i > 0, \ i = 1, \ldots, n.$$

Physical interpretation of the effect of input

The material throughput flow induces nondissipative contributions to the system by adding an entropy flux term to the entropy balance. In our formalism, this balance is obtained by computing the time derivative of S, as defined in (7.114) along (7.116):

$$\dot{S} = (\mu - \mu^*)^T \sum_{j=1}^{r} \nu_j W_j + (\mu - \mu^*)^T \phi(x^0 - x). \tag{7.119}$$

The second term on the right-hand side of (7.119) corresponds to the entropy flux and may compensate for or even override the natural entropy dissipation, thus undermining the inherent global asymptotic stability of the system. At this point, the direct relationship between entropy flux and dynamical complexity can be noted. Therefore, in stabilizing open complex reaction systems, it seems crucial to act on the nondissipative contributions by appropriate control configurations.

7.5.5 Local Hamiltonian Description of Reversible Reaction Networks

System descriptions derived from a potential form an interesting class of dynamic systems where powerful methods for control design, such as those based on passivity, apply directly. In this section, we demonstrate that a complex reaction network possesses an underlying potential structure on a state-space that will be referred to as the *reaction space*.

Consider a reversible reaction network with no external input described in (7.105). The reaction space variables are obtained by the following nonlinear coordinate transformation:

$$z_j = \ln \frac{p_j}{q_j}, \quad j = 1, \ldots, r, \tag{7.120}$$

where $p_j = W_j^+$ and $q_j = W_j^-$ are the direct and reverse rates associated with the reaction rate. In the new variables, (7.103) becomes

$$W_j = p_j - q_j = q_j(e^{z_j} - 1). \tag{7.121}$$

The right-hand side of (7.112) can then be transformed through appropriate manipulations into the form,

$$\ell(z, q) = -\sum_{j=1}^{r} z_j \, q_j(e^{z_j} - 1) = -z^T W. \tag{7.122}$$

Function ℓ can be easily connected with the so-called dissipation function because it is the product of thermodynamic fluxes (reaction rates) and thermodynamic forces (chemical affinities). In this way, it seems natural to explore the properties of chemical reaction network dynamics in the *reaction space* defined by z variables. For that purpose, let us introduce the following notations:

$$F(q) = \text{diag}\{q_1, \cdots, q_r\}, \tag{7.123}$$

$$S(x) = \mathcal{N}^T \Gamma(x) \mathcal{N}, \tag{7.124}$$

where

$$\Gamma(x) = \text{diag}\left\{\frac{1}{x_1}, \cdots, \frac{1}{x_i}, \cdots \frac{1}{x_n}\right\}. \tag{7.125}$$

The time derivative of z_j can be calculated as

$$\dot{z}_j = \frac{1}{p_j}\dot{p}_j - \frac{1}{q_j}\dot{q}_j = \frac{1}{p_j}\frac{\partial p_j}{\partial x}\dot{x} - \frac{1}{q_j}\frac{\partial q_j}{\partial x}\dot{x}$$

$$= \left[\frac{1}{x_1}(\alpha_{1j} - \beta_{1j}) \quad \frac{1}{x_2}(\alpha_{2j} - \beta_{2j}) \quad \cdots \quad \frac{1}{x_n}(\alpha_{nj} - \beta_{nj})\right]\dot{x}. \tag{7.126}$$

This means that $\dot{z}$ is given by:

$$\dot{z} = -\mathcal{N}^T \Gamma(x) \cdot \dot{x} = -\mathcal{N}^T \Gamma(x) \mathcal{N} \cdot W(x). \tag{7.127}$$

Let us define the Hamiltonian function in the following form:

$$H(z) = \sum_{i=1}^{r} q_i^* \left[\exp(z_i) - z_i - 1 \right], \tag{7.128}$$

where q_i^* denotes the value of q_i at the equilibrium point x^*. It is easy to show that H is globally convex and bounded from below. Therefore, it can be used as a Lyapunov function.

Using (7.121), (7.127) and (7.128), we can write

$$\dot{z} = -\mathcal{N}^T \Gamma(x) \mathcal{N} \cdot F(q) \cdot [F(q^*)]^{-1} \cdot \frac{\partial H}{\partial z}(z), \tag{7.129}$$

or briefly,

$$\dot{z} = -G(x) \cdot \frac{\partial H}{\partial z}(z), \tag{7.130}$$

where

$$G(x) = \mathcal{N}^T \Gamma(x) \mathcal{N} \cdot F(q) \cdot [F(q^*)]^{-1}. \tag{7.131}$$

It is easy to see that G is positive definite in the neighbourhood of the equilibrium point, since

$$G(x^*) = \mathcal{N}^T \Gamma(x^*) \mathcal{N} \cdot F(q^*) \cdot [F(q^*)]^{-1} = \mathcal{N}^T \Gamma(x^*) \mathcal{N}. \tag{7.132}$$

However, it cannot be guaranteed that G is positive definite in the whole concentration space for an arbitrary reaction network. Therefore, we have shown that reversible reaction networks have a *local dissipative Hamiltonian structure* in transformed z coordinates (the reaction space).

Example

Consider the following elementary reaction network with three species P_1, $P2$ and P_3:

$$P_1 \underset{k_2}{\overset{k_1}{\rightleftarrows}} P_2 \underset{k_4}{\overset{k_3}{\rightleftarrows}} P_3. \tag{7.133}$$

Let us denote the concentrations of P_1, P_2 and P_3 by x_1, x_2 and x_3, respectively. Then, the structure matrix $\mathcal{N}$ and the reaction rate vector $W(x)$ can be written as

$$\mathcal{N} = \begin{bmatrix} -1 & 0 \\ 1 & -1 \\ 0 & 1 \end{bmatrix}, \tag{7.134}$$

$$W(x) = \begin{bmatrix} k_1 x_1 - k_2 x_2 \\ k_3 x_2 - k_4 x_3 \end{bmatrix}. \tag{7.135}$$

For the sake of simplicity, let us choose all reaction rate constants as 1, *i.e.*, $k_1 = k_2 = k_3 = k_4 = 1$. It is easy to calculate (using *e.g.*, the kernel of N^T) that the following conservation law is valid for the system:

$$x_1(t) + x_2(t) + x_3(t) = x_1(0) + x_2(0) + x_3(0) = x_0, \quad \forall\, t > 0. \tag{7.136}$$

Furthermore, it can be seen from W that the set of equilibrium points is located on the following subspace:

$$X^* = \{x \mid x_1 = x_2 = x_3\}. \tag{7.137}$$

The transformed z coordinates of the reaction space are computed as

$$z_1 = \ln\left(\frac{p_1}{q_1}\right) = \ln\left(\frac{x_1}{x_2}\right),$$

$$z_2 = \ln\left(\frac{p_2}{q_2}\right) = \ln\left(\frac{x_2}{x_3}\right).$$

Since $q_i^* = 1$ for $i = 1, 2$, the Hamiltonian function for the system is given by

$$H(z) = \sum_{i=1}^{2} \left[e^{z_i} - z_i - 1\right]. \tag{7.138}$$

Let us choose the particular equilibrium point $x^* = [1\ 1\ 1]^T$ from X^* for studying the local dissipative Hamiltonian description of the network.

The matrix $G(x)$ is given by

$$G(x) = \begin{bmatrix} x_2\left(\frac{1}{x_1} + \frac{1}{x_2}\right) & -\frac{x_3}{x_2} \\ -1 & x_3\left(\frac{1}{x_2} + \frac{1}{x_3}\right) \end{bmatrix}. \tag{7.139}$$

To study the definiteness of G, we need its symmetrized matrix which (using the conservation law (7.136)) can be written as,

$$G_s(x) = \frac{1}{2}\left[G(x) + G^T(x)\right]$$

$$= \frac{1}{2}\begin{bmatrix} 2x_2\left(\frac{1}{x_1} + \frac{1}{x_2}\right) & 7 - 1 - \frac{3 - x_1 - x_2}{x_2} \\ -1 - \frac{3 - x_1 - x_2}{x_2} & 2(3 - x_1 - x_2)\left(\frac{1}{x_2} + \frac{1}{3 - x_1 - x_2}\right) \end{bmatrix}. \tag{7.140}$$

The value of G_s at the equilibrium point is

$$G_s(x^*) = \begin{bmatrix} 2 & -1 \\ -1 & 2 \end{bmatrix}, \tag{7.141}$$

which is clearly positive definite.

In general, the first minor of G_s is always positive, and its determinant can be computed as

$$\det(G_s(x)) = -\frac{(4x_2^2 + 4x_1x_2 - 3x_1 + x_1^2)(-3 + x_1)}{4x_2^2 x_1}.$$ (7.142)

It can be shown that $\det(G_s(x))$ is positive in a wide neighbourhood of the equilibrium point, but not in the whole concentration space. This means that the generalized dissipative Hamiltonian description (7.129) is valid in the neighbourhood of the equilibrium.

7.6 Summary

The thermodynamic foundations of passivity-based process control described in this chapter include the postulates and laws of classical thermodynamics, the Onsager relationship from nonequilibrium thermodynamics and the various sets of canonical thermodynamic variables. It was shown that thermodynamics has strong implications for the structure of the state equations of process systems by determining the number and kind of state variables and the type of nonlinearities.

A thermodynamically motivated entropy-based storage function was also proposed that can be applied in the case of commonly chosen input and output variables, where the input variables are the intensive variables at the inlets.

The Hamiltonian description of a process system was also introduced. It offers another possibility of finding physically motivated storage functions for such cases when the mass inflow/outflow rate is chosen as an input variable.

Acknowledgement

The support of the Hungarian Scientific Research Fund through grants T042710, F046223 is gratefully acknowledged. The second author acknowledges the support of the János Bolyai Research Scholarship of the Hungarian Academy of Sciences.

A

Detailed Control Design Algorithms

A.1 Solution to the BMI Problem in SPR/$\mathcal{H}_\infty$ Control Design

The SPR/$\mathcal{H}_\infty$ controller synthesis problem formulated in Section 3.5 requires the solution of the following inequalities:

$$\begin{bmatrix} I & Q^T \\ Q & I \end{bmatrix} > 0, \tag{A.1}$$

$$\Psi_k + Z^T Q^T P_k + P_k^T Q Z < 0. \tag{A.2}$$

This is a BMI problem because (A.2) has bilinear terms in two decision variables: Q and P. BMI problems are not convex and NP-hard. Here we present an approach developed in [148, 150]. The basic idea is to start with a central $\mathcal{H}_\infty$ controller ($Q = 0$) and find a solution path that leads to SPR/$\mathcal{H}_\infty$ conditions. Because the initial central $\mathcal{H}_\infty$ controller usually does not satisfy the SPR condition, (A.2) is replaced by a relaxed condition:

$$\begin{bmatrix} A_k^T P + P A_k & P B_k - C_k^T \\ B_k^T P - C_k & -D_k - D_k^T \end{bmatrix} < \alpha \begin{bmatrix} P & 0 \\ 0 & Y \end{bmatrix}, \tag{A.3}$$

or equivalently,

$$\begin{bmatrix} T_1 & T_2 \\ T_3 & T_4 \end{bmatrix} < 0, \tag{A.4}$$

where

$$T_1 = A_F^T P + P A_F + C_{2F}^T Q^T B_{2F}^T P + P B_{2F} Q C_{2F} - \alpha P,$$
$$T_2 = P B_{1F} + P B_{2F} Q D_{21F} - C_{1F}^T - C_{2F}^T Q^T D_{12F}^T,$$
$$T_3 = B_{1F}^T P + D_{21F}^T Q^T B_{2F}^T P - C_{1F} - D_{12F} Q C_{2F} \text{ and}$$
$$T_4 = -D_{11F} - D_{11F}^T - D_{12F} Q D_{21F} - D_{21F}^T Q^T D_{12F}^T - \alpha Y.$$

$P > 0, Y > 0$ and α is a scalar decision variable. When $\alpha < 0$, the relaxed condition recovers to the original SPR condition. Therefore, minimizing α subject to (A.3) leads to the SPR condition. The minimization problem is a generalized eigenvalue problem on P, Y and α and can be solved using semidefinite programming (SDP) techniques. Note that minimizing $[\mathrm{Tr}(P) + \mathrm{Tr}(Y)]$ is essential for the convergence of α to a local minimum [27]. Therefore, for the given initial $\mathcal{H}_\infty$ controller, minimizations of α and $\mathrm{Tr}(P) + \mathrm{Tr}(Y)$ are implemented alternately to get the smallest α and corresponding P and Y. P and Y obtained are then used to find a new Q by solving another minimization problem for α, subject to (A.4) and (A.1). Iterations continue until $\alpha < 0$, which gives a feasible solution of the SPR/$\mathcal{H}_\infty$ control problem.

A.2 DUS $\mathcal{H}_2$ Control Synthesis

This algorithm was developed by the authors and their co-worker and first published in [16].

A.2.1 Final LMI

By approximating the bilinear constraints using (5.63), and partitioning the symmetric matrices P_2 and P_{20} into the following forms:

$$P_2 = \begin{bmatrix} P_{211} & P_{212} & P_{213} \\ P_{212}^T & P_{222} & P_{223} \\ P_{213}^T & P_{223}^T & P_{233} \end{bmatrix} \text{ and } P_{20} = \begin{bmatrix} P_{2110} & P_{2120} & P_{2130} \\ P_{2120}^T & P_{2220} & P_{2230} \\ P_{2130}^T & P_{2230}^T & P_{2330} \end{bmatrix}. \tag{A.5}$$

Problem 5.12 is converted into the following problem:

Problem A.1 ([16]).

$$\min_{\delta K_{gi}, \delta k_{si}, \delta P_1, \delta P_2, \delta Q} \{\mathrm{Tr}(\delta Q)\}, \tag{A.6}$$

subject to

$$\begin{bmatrix} S & \delta P_1^T L - \delta K_{gi}^T + P_{10}^T L - K_{gi0}^T \\ L^T \delta P_1 - \delta K_{gi} + L^T P_{10} - K_{gi0} & 0 \end{bmatrix} \leq 0, \tag{A.7}$$

$$k_{si0} + \delta k_{si} > 0, \tag{A.8}$$

$$P_{10} + \delta P_1 > 0, \tag{A.9}$$

$$\begin{bmatrix} Z_{11} & Z_{21}^T & Z_{31}^T & Z_{41}^T \\ Z_{21} & Z_{22} & Z_{32}^T & Z_{42}^T \\ Z_{31} & Z_{32} & Z_{33} & Z_{43}^T \\ Z_{41} & Z_{42} & Z_{43} & Z_{44} \end{bmatrix} < 0, \tag{A.10}$$

$$\begin{bmatrix} P_{2110} + \delta P_{211} & P_{2120} + \delta P_{212} & P_{2130} + \delta P_{213} & C_1^T \\ P_{2120}^T + \delta P_{212}^T & P_{2220} + \delta P_{222} & P_{2230} + \delta P_{223} & K_{gi0}^T D_{12}^T + \delta K_{gi}^T D_{12}^T \\ P_{2130}^T + \delta P_{213}^T & P_{2230}^T + \delta P_{223}^T & P_{2330} + \delta P_{233} & k_{si0}^T D_{12}^T + \delta k_{si}^T D_{12}^T \\ C_1 & D_{12} K_{gi0} + D_{12} \delta K_{gi} & D_{12} k_{si0} + D_{12} \delta k_{si} & Q_0 + \delta Q \end{bmatrix} > 0,$$

(A.11)

where

$$\begin{aligned} S =& P_{10}E + \delta P_1 E + P_{10}FK_{gi0} + P_{10}F\delta K_{gi} + \delta P_1 FK_{gi0} + E^T P_{10}^T \\ &+ E^T \delta P_1^T + K_{gi0}^T F^T P_{10}^T + \delta K_{gi}^T F^T P_{10}^T + K_{gi0}^T F^T \delta P_1^T, \end{aligned} \qquad \text{(A.12)}$$

$$E = A - LC_2, \qquad \text{(A.13)}$$

$$F = B_2 - LD_{22}, \qquad \text{(A.14)}$$

$$\begin{aligned} Z_{11} =& P_{2110}A + \delta P_{211}A + P_{2120}LC_2 + \delta P_{212}LC_2 + P_{2130}C_2 \\ &+ \delta P_{213}C_2 + A^T P_{2110}^T + A^T \delta P_{211}^T + C_2^T L^T P_{2120}^T + C_2^T L^T \delta P_{212}^T \\ &+ C_2^T P_{2130}^T + C_2^T \delta P_{213}^T, \end{aligned} \qquad \text{(A.15)}$$

$$\begin{aligned} Z_{21} =& K_{gi0}^T B_2^T P_{2110}^T + \delta K_{gi}^T B_2^T P_{2110}^T + K_{gi0}^T B_2^T \delta P_{211}^T + A^T P_{2120}^T \\ &+ A^T \delta P_{212}^T + K_{gi0}^T B_2^T P_{2120}^T + \delta K_{gi}^T B_2^T P_{2120}^T + K_{gi0}^T B_2^T \delta P_{212}^T \\ &- C_2^T L^T P_{2120}^T - C_2^T L^T \delta P_{212}^T + P_{2220}LC_2 + \delta P_{222}LC_2 \\ &+ P_{2120}^T A + \delta P_{212}^T A + P_{2230}C_2 + \delta P_{223}C_2 + K_{gi0}^T D_{22}^T P_{2130}^T \\ &+ \delta K_{gi}^T D_{22}^T P_{2130}^T + K_{gi0}^T D_{22}^T \delta P_{213}^T \end{aligned} \qquad \text{(A.16)}$$

$$\begin{aligned} Z_{22} =& P_{2120}^T B_2 K_{gi0} + P_{2120}^T B_2 \delta K_{gi} + \delta P_{212}^T B_2 K_{gi0} + P_{2220}A \\ &+ \delta P_{222}A + P_{2220}B_2 K_{gi0} + P_{2220}B_2 \delta K_{gi} + \delta P_{222}B_2 K_{gi0} \\ &- P_{2220}LC_2 - \delta P_{222}LC_2 + P_{2230}D_{22}K_{gi0} + \delta P_{223}D_{22}K_{gi0} \\ &+ P_{2230}D_{22}\delta K_{gi} + K_{gi0}^T B_2^T P_{2120} + \delta K_{gi}^T B_2^T P_{2120} \\ &+ K_{gi0}^T B_2^T \delta P_{212} + A^T P_{2220}^T + A^T \delta P_{222}^T + K_{gi0}^T B_2^T P_{2220}^T \\ &+ \delta K_{gi}^T B_2^T P_{2220}^T + K_{gi0}^T B_2^T \delta P_{222}^T - C_2^T L^T P_{2220}^T - C_2^T L^T \delta P_{222}^T \\ &+ K_{gi0}^T D_{22}^T P_{2230}^T + \delta K_{gi}^T D_{22}^T P_{2230}^T + K_{gi0}^T D_{22}^T \delta P_{223}^T \end{aligned} \qquad \text{(A.17)}$$

$$\begin{aligned} Z_{31} =& P_{2130}^T A + \delta P_{213}^T A + P_{2230}^T LC_2 + \delta P_{223}^T LC_2 + P_{2330}C_2 \\ &+ \delta P_{233}C_2 + k_{si0}B_2^T P_{2110}^T + \delta k_{si}B_2^T P_{2110}^T + k_{si0}B_2^T \delta P_{211}^T \\ &+ k_{si0}D_{22}^T L^T P_{2120}^T + \delta k_{si}D_{22}^T L^T P_{2120}^T + k_{si0}D_{22}^T L^T \delta P_{212}^T \\ &+ k_{si0}D_{22}^T P_{2130}^T + \delta k_{si}D_{22}^T P_{2130}^T + k_{si0}D_{22}^T \delta P_{213}^T, \end{aligned} \qquad \text{(A.18)}$$

$$
\begin{aligned}
Z_{32} =\ & P_{2130}^{T} B_2 K_{gi0} + \delta P_{213}^{T} B_2 K_{gi0} + P_{2130}^{T} B_2 \delta K_{gi} + P_{2230}^{T} A \\
& + \delta P_{223}^{T} A + P_{2230}^{T} B_2 K_{gi0} + \delta P_{223}^{T} B_2 K_{gi0} + P_{2230}^{T} B_2 \delta K_{gi} \\
& - P_{2230}^{T} L C_2 - \delta P_{223}^{T} L C_2 + P_{2230}^{T} D_{22} K_{gi0} + \delta P_{223}^{T} D_{22} K_{gi0} \\
& + P_{2230}^{T} D_{22} \delta K_{gi} + k_{si0} B_2^{T} P_{2120} + \delta k_{si} B_2^{T} P_{2120} + k_{si0} B_2^{T} \delta P_{212} \\
& + k_{si0} D_{22}^{T} L^{T} P_{2220}^{T} + \delta k_{si} D_{22}^{T} L^{T} P_{2220}^{T} + k_{si0} D_{22}^{T} L^{T} \delta P_{222}^{T} \\
& + k_{si0} D_{22}^{T} P_{2230}^{T} + \delta k_{si} D_{22}^{T} P_{2230}^{T} + k_{si0} D_{22}^{T} \delta P_{223}^{T},
\end{aligned} \tag{A.19}
$$

$$
\begin{aligned}
Z_{33} =\ & P_{2130}^{T} B_2 k_{si0} + \delta P_{213}^{T} B_2 k_{si0} + P_{2130}^{T} B_2 \delta k_{si} + P_{2230}^{T} L D_{22} k_{si0} \\
& + \delta P_{223}^{T} L D_{22} k_{si0} + P_{2230}^{T} L D_{22} \delta k_{si} + P_{2330} D_{22} k_{si0} + \delta P_{233} D_{22} k_{si0} \\
& + P_{2330} D_{22} \delta k_{si} + k_{si0} B_2^{T} P_{2130} + \delta k_{si} B_2^{T} P_{2130} + k_{si0} B_2^{T} \delta P_{213} \\
& + k_{si0} D_{22}^{T} L^{T} P_{2230} + \delta k_{si} D_{22}^{T} L^{T} P_{2230} + k_{si0} D_{22}^{T} L^{T} \delta P_{223} \\
& + k_{si0} D_{22}^{T} P_{2330}^{T} + \delta k_{si} D_{22}^{T} P_{2330}^{T} + k_{si0} D_{22}^{T} \delta P_{233}^{T},
\end{aligned} \tag{A.20}
$$

$$
\begin{aligned}
Z_{41} =\ & B_1^{T} P_{2110}^{T} + B_1^{T} \delta P_{211}^{T} + D_{21}^{T} L^{T} P_{2120}^{T} + D_{21}^{T} L^{T} \delta P_{212}^{T} + D_{21}^{T} P_{2130}^{T} \\
& + D_{21}^{T} \delta P_{213}^{T},
\end{aligned} \tag{A.21}
$$

$$
\begin{aligned}
Z_{42} =\ & B_1^{T} P_{2120} + B_1^{T} \delta P_{212} + D_{21}^{T} L^{T} P_{2220}^{T} + D_{21}^{T} L^{T} \delta P_{222}^{T} + D_{21}^{T} P_{2230}^{T} \\
& + D_{21}^{T} \delta P_{223}^{T},
\end{aligned} \tag{A.22}
$$

$$
\begin{aligned}
Z_{43} =\ & B_1^{T} P_{2130} + B_1^{T} \delta P_{213} + D_{21}^{T} L^{T} P_{2230} + D_{21}^{T} L^{T} \delta P_{223} + D_{21}^{T} P_{2330}^{T} \\
& + D_{21}^{T} \delta P_{233}^{T} \text{ and}
\end{aligned} \tag{A.23}
$$

$$
Z_{44} = -I. \tag{A.24}
$$

A.2.2 SSDP Procedure

The following procedure is implemented in designing each controller loop $k_i(s)$ $(i = 1, \ldots, m)$.

Procedure A.2 ([16])

1. *Find the initial set of solutions for K_{gi0}, k_{si0}, P_{10}, P_{20} and Q_0 using the method described above.*
2. *Set the solution radius $\epsilon = \epsilon_0$. It should be set at a small positive number so that the optimization solver in Step 3 will search in the small neighbourhood of initial values. Also set convergence tolerance ζ and the maximum number of iterations η.*
3. *Check whether the maximum number of iterations has been reached. If yes, terminate the SSDP procedure and stop.*
4. *Solve Problem A.1 (the problem with approximated constraints) with restrictions on the solution radii:*

$$
\|\delta K_{gi}\| < \epsilon,\ \delta k_{si} < \epsilon, \|\delta P_1\| < \epsilon, \|\delta P_2\| < \epsilon, \|\delta Q\| < \epsilon, \tag{A.25}
$$

to obtain δK_{gi}, δk_{si}, δP_1, δP_2 and δQ. Because zero initial values of decision variables (deviation variables) satisfy all constraints in Problem A.1, there always exists a feasible solution.

5. Update $K_{gi} = K_{gi0} + \delta K_{gi}$ and $k_{si} = k_{si0} + \delta k_{si}$. Fix K_{gi} and k_{si}, and solve Problem A.1 with the LMIs for P_1, P_2 and Q. This step provides the solution of the original BMI problem.

a) If no feasible solution is obtained, the solution radius in Step 4 is too large and the solution of the approximated problem does not satisfy the original nonlinear constraints. Reduce the solution radius by replacing ϵ with 0.9ϵ and go to Step 3.

b) If a feasible solution is obtained, there are two scenarios:

i. If $\text{Tr}(Q) > \text{Tr}(Q_0)$, the new solution is less optimal than the previous solution. Increase the solution radius by replacing ϵ with 1.1ϵ and go to Step 3.

ii. If $\text{Tr}(Q) < \text{Tr}(Q_0)$ with the new solution, a better result is obtained. Now check for convergence:

A. If $|\text{Tr}(Q) - \text{Tr}(Q_0)| < \zeta$, then an acceptable solution is obtained and proceed to Step 6.

B. Otherwise, set $K_{gi0} = K_{gi}$, $k_{si0} = k_{si}$, $P_{10} = P_1$, $P_{20} = P_2$ and $Q_0 = Q$. Go to Step 3.

6. Calculate $k_i'(s)$ using (5.52) and the final controller $k_i(s)$ is computed using (5.53).

B

Mathematical Proofs

B.1 Phase Condition for MIMO Systems

To derive the phase condition for MIMO systems, we need to use the concept of inertia.

Definition B.1 (Inertia of a matrix[91]). *For an $m \times m$ matrix A, $\text{In}(A) = (\pi, \nu, \sigma)$ is denoted as its inertia, which means that matrix A has π eigenvalues with positive real parts, ν eigenvalues with negative real parts and σ purely imaginary eigenvalues.*

Theorem B.2 ([91]). *If for an $m \times m$ matrix A, $(A + A^*) > 0$, and H is Hermitian, then*

$$\text{In}(AH) = \text{In}(H). \tag{B.1}$$

Lemma B.3 ([91]). *Let A be an $m \times m$ matrix and H be Hermitian. If $AH + H^*A^*$ is positive definite, then H is nonsingular.*

To simplify our proof, let us look at strictly input passive systems.

Theorem B.4 ([13]). *Consider an $m \times m$ MIMO LTI system with a transfer function $G(s)$. If the system is strictly passive, then its phase shift lies in the open interval $(-90°, 90°)$ for any real ω.*

Proof. If system $G(s)$ is strictly passive, the following inequality holds by Theorem 2.25:

$$G(j\omega) + G^*(j\omega) = U(j\omega)H(j\omega) + H^*(j\omega)U^*(j\omega) > 0. \tag{B.2}$$

From Lemma B.3, the Hermitian matrix $H(j\omega)$ is nonsingular. Furthermore, since

$$\lambda\left(H(j\omega)\right) = \lambda\left(V(j\omega)\Lambda(j\omega)V^*(j\omega)\right) \tag{B.3a}$$

$$= \lambda\left(V(j\omega)\Lambda(j\omega)V^{-1}(j\omega)\right) = \lambda\left(\Lambda(j\omega)\right), \tag{B.3b}$$

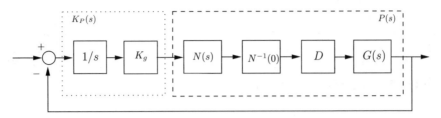

Fig. B.1. Passivity-based DIC [12]

we can conclude that all eigenvalues of $H(j\omega)$ are positive. Because H^{-1} is also a positive definite Hermitian matrix and In $\left(H^{-1}(j\omega)\right) = (m, 0, 0)$, from Theorem B.2,

$$\text{In}\left(U(j\omega)\right) = \text{In}\left([U(j\omega)H(j\omega)]H^{-1}\right) = \text{In}\left(H^{-1}(j\omega)\right) = (m, 0, 0), \quad (\text{B.4})$$

that is, all eigenvalues of $U(j\omega)$ have positive real parts. Consequently,

$$\arg\left\{\lambda_i\left(U(j\omega)\right)\right\} \in (-90°, 90°) \tag{B.5}$$

for all $\omega \in \mathbb{R}$, $i = 1\ldots m$, where the principal argument is defined in $(-180°, 180°)$.

B.2 Proof of Theorem 4.4

Proof ([12]). Consider the system shown in Figure B.1. Assume that a decentralized controller with integral action $C(s)$ is formed by

$$C(s) = DN^{-1}(0)N(s)K_g/s = DN^{-1}(0)N(s)K_p(s), \tag{B.6}$$

where $K_g = \text{diag}\{k_i\}$, $k_i \geq 0, k_i \in \mathbb{R}$, $i = 1\ldots m$, $K_p(s) = K_g/s$; $N(s)$ is a stable, diagonal system and $N(0)$ is nonsingular. From Definition 2.24, $K_p(s)$ is positive real. Denote $P(s) = G(s)DN^{-1}(0)N(s)$. The DIC condition transforms to a determination of whether the closed-loop system of $K_p(s)$ and $P(s)$ is stable and remains stable when K_g is reduced to $K_{ge} = \text{diag}\{k_i\varepsilon_i\}$ $(0 \leq \varepsilon_i \leq 1, i = 1\ldots m)$. The stability is proven using the *generalized nyquist stability criterion*. Since $K_p(s)$ has nonrepeated poles at the origin $(s = 0)$, the standard D-contour has to be modified by making a semicircular "detour" with infinitesimal radius ε $(\varepsilon \to 0)$ to the right of the origin. Define the points on the semicircle, the rest of the imaginary axis and the right half circle with infinite radius as S_ε, S_i, S_r respectively, where

$$S_\varepsilon = \left\{s : |s = \varepsilon e^{j\theta}, \varepsilon \in R^+, \varepsilon \to 0, \text{ and } -\pi/2 < \theta < \pi/2\right\},$$

$$S_i = \left\{s : |s = j\omega, \omega = (-\infty, -\varepsilon] \cup [+\varepsilon, +\infty), \varepsilon \in R^+, \varepsilon \to 0\right\} \text{ and}$$

$$S_r = \left\{s : |s = re^{j\theta}, r \in R^+, r \to +\infty \text{ and } -\pi/2 < \theta < \pi/2\right\}.$$

Because both $P(s)$ and $K_p(s)$ do not have RHP poles, the *generalized nyquist stability criterion* reduces to the condition that the eigenvalue loci of $P(s)K_p(s)$ do not encircle $s = -1 + 0j$ for any $s \in S_\varepsilon$ and for any $s \in S_i \cup S_r$.

Because both $N(0)$ and $G(0)$ are nonsingular, if there exists a matrix D such that (4.11) is satisfied (*i.e.*, $P(0) + P^T(0) \geq 0$), then $P(0)$ is nonsingular and

$$\mathrm{Re}\left[z^* P(0) z\right] \geq 0 \quad \text{and} \quad \mathrm{Re}\left[z^* P^T(0) z\right] \geq 0 \ \forall z \neq 0, \ z \in \mathbb{C}^m. \tag{B.7}$$

Proof part A: For any $s \in S_\varepsilon$,

$$K_p(s) + K_p^*(s) = \frac{K_g}{\varepsilon e^{j\theta}} + \frac{K_g^T}{\varepsilon e^{-j\theta}} = \frac{2K_g \cos(\theta)}{\varepsilon} \geq 0. \tag{B.8}$$

Therefore,

$$\mathrm{Re}\left(z^*\left[K_p(s)\right] z\right) = \mathrm{Re}\left(z^*\left[K_p^*(s)\right] z\right) \geq 0 \ \forall \ z \neq 0, \ z \in \mathbb{C}^m \text{ and } s \in S_\varepsilon. \tag{B.9}$$

Since $G(s)$ and $N(s)$ are stable, therefore $P(s)$ is analytic and thus continuous in $\mathrm{Re}(s) \geq 0$,

$$\lim_{\varepsilon \to 0} P\left(\varepsilon e^{j\theta}\right) = P(0), \tag{B.10}$$

where $-\pi/2 < \theta < \pi/2$. From (B.7), for any $s \in S_\varepsilon$, $P(s)$ is nonsingular and

$$\mathrm{Re}\left[z^* P(s) z\right] = \lim_{\varepsilon \to 0} \mathrm{Re}\left[z^* P(\varepsilon e^{j\theta}) z\right]$$
$$= \mathrm{Re}\left[z^* P(0) z\right] \geq 0 \text{ for any } z \neq 0, \ -\pi/2 < \theta < \pi/2. \tag{B.11}$$

If there exists a real eigenvalue λ of $P(s)K_p(s)$, $s \in S_\varepsilon$, it is also an eigenvalue of $K_p(s)P(s)$. Assuming that $x \neq 0$ is the eigenvector associated with the eigenvalue λ of $K_p(s)P(s)$ (*i.e.*, $\lambda x = K_p(s)P(s)x$),

$$\mathrm{Re}[\lambda x^* P(s)x] = \mathrm{Re}[x^* P^*(s)K_p^*(s)P(s)x]$$
$$= \mathrm{Re}[y^* K_p^*(s)y], \quad \forall s \in S_\varepsilon, \tag{B.12}$$

where $y = P(s)x \neq 0$. From (B.9),

$$\mathrm{Re}\left[\lambda x^* P(s)x\right] = \lambda \mathrm{Re}\left[x^* P(s)x\right] = \mathrm{Re}[y^* K_p^*(s)y] \geq 0. \tag{B.13}$$

From (B.11),

$$\mathrm{Re}\left[x^* P(s)x\right] \geq 0 \tag{B.14}$$

for the eigenvector x defined in (B.12) and $s \in S_\varepsilon$.

(a) For the case in which $\mathrm{Re}\left[x^* P(s)x\right] > 0$, $s \in S_\varepsilon$, because the eigenvalue λ is real, from (B.13),

$$\lambda \geq 0. \tag{B.15}$$

(b) For the case in which $\mathrm{Re}\,[x^* P(s)x] = 0$, it can be derived from (B.12) that

$$\mathrm{Re}[y^* K_p(s)y] = \mathrm{Re}[y^* K_p^*(s)y] = 0 \text{ for any } s \in S_\varepsilon. \tag{B.16}$$

Recalling $K_p(s) = \mathrm{diag}\,\{k_{p,1}, \ldots, k_{p,i}, \ldots, k_{p,m}\}$, as defined in (B.6) and $y = [y_1, \ldots y_i, \ldots, y_n]^T$ (where $k_{p,i},\ y_i \in \mathbb{C}^1\ \forall\ i = 1, \ldots, m$),

$$\mathrm{Re}[y^* K_p(s)y] = \sum_{i=1}^{m} \mathrm{Re}\,(k_{p,i})\,\overline{y}_i y_i = 0, \tag{B.17}$$

where $\overline{y}$ is the complex conjugate of y.

Inequality B.8 implies that $\mathrm{Re}\,(k_{p,i}) \geq 0$, since $\overline{y}_i y_i$ is a real number and $\overline{y}_i y_i \geq 0$. It can be seen from (B.17) that,

$$\mathrm{Re}\,(k_{p,i}) = 0 \text{ if } \overline{y}_i y_i \neq 0. \tag{B.18}$$

Because $k_i \geq 0$, $k_i \in \mathbb{R}$, and $\mathrm{Re}\left(\varepsilon e^{j\theta}\right) > 0$,

$$k_{p,i} = 0 \ \forall\ k_{p,i} \in \{k_{p,i}\,|\mathrm{Re}\,(k_{p,i}) = 0\}\ ,\quad i = 1, \ldots, m. \tag{B.19}$$

Then,

$$\lambda x = K_p(s)\,P(s)\,x = 0,\quad x \neq 0. \tag{B.20}$$

This leads to

$$\lambda = 0. \tag{B.21}$$

Therefore, any real eigenvalue of $K_p(s)\,P(s)$ is always greater than or equal to zero. This implies that the eigenvalue locus of $P(s)K_p(s)$ does not cross the negative real axis and thus does not encircle the critical point $s = -1 + 0j$, while s traverses on the semicircle.

Proof part B: For any $s \in S_i \cup S_r$: Since $P(s)$ is analytic in $\mathrm{Re}\,(s) \geq 0$ and

$$P(s)K_p(s) = G(s)DN^{-1}(0)\,N(s)K_g s^{-1} = G(s)K(s) \tag{B.22}$$

is proper, there always exists a positive definite diagonal matrix $K_s = \mathrm{diag}\{k_{s,i}\}$, $k_{s,i} > 0$, $i = 1 \ldots m$, with its elements small enough such that the following condition is satisfied:

$$\max_{s \in S_i \cup S_r} \left|\lambda_{\max}\left(P(s)K_s \frac{1}{s}\right)\right| \leq \sigma_{\max}\,(P(s))\,k_{sm}\left|\frac{1}{s}\right| < 1, \tag{B.23}$$

where k_{sm} is the maximum element of K_s and $\sigma_{\max}\,(P(s))$ denotes the maximum singular value of $P(s)$.

Therefore, by choosing any $K_g \geq 0$, whose maximum element is less than k_{sm}, the moduli of the eigenvalues of $P(s)K_p(s)$ are always less than 1 as s travels on the imaginary axis from $-\infty$ to $-\varepsilon$ and from $+\varepsilon$ to $+\infty$.

By combining Proof part A and part B, it is concluded that it is always possible to find a K_g such that the eigenvalue locus of $P(s)K_p(s)$ will never

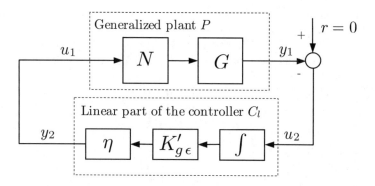

Fig. B.2. Nonlinear DIC analysis

encircle the critical point $(-1, 0j)$ when s travels on the D-contour. The closed-loop stability of $P(s)$ and $K_p(s)$ is concluded.

It is obvious that if K_g is reduced to any positive semidefinite diagonal matrix,

$$K_{g\varepsilon} = \mathrm{diag}\{k_i\varepsilon_i\}, \ 0 \le \varepsilon_i \le 1, \ i = 1 \ldots m, \tag{B.24}$$

the above stability condition is still valid.

B.3 Proof of Theorem 4.8

Proof ([119]). The proof of this theorem is based on singular perturbation theory [70]. Consider the system of Figure B.2 described by (4.24) and (4.25). The input signal r is set to zero so that the Lyapunov stability of the unforced closed loop $(P, -C_l)$ can be analyzed. The state equation for the closed loop $(P, -C_l)$ can be expressed as

$$(P, -C_l): \ \begin{cases} \dot{x} = f(x, \xi) \\ \dot{\xi} = -\eta K'_{g\varepsilon} g(x, \xi). \end{cases} \tag{B.25}$$

Equation B.25 can be transformed into a standard singular perturbation form [70]: Let $\tau = \eta(t - t_0)$, so that $\tau = 0$ at $t = t_0$. Because $\frac{d\tau}{dt} = \eta$,

$$(P, -C_l): \ \begin{cases} \eta \frac{d}{d\tau} x = f(x, \xi) \\ \frac{d}{d\tau} \xi = -K'_{g\varepsilon} g(x, \xi). \end{cases} \tag{B.26}$$

To be consistent with standard singular perturbation notation, we will for the moment use the notation $\dot{x}$ to denote the derivative in the *slow* timescale τ when we analyse singular perturbation models. For the standard singular perturbation model (B.26), the following conclusions can be drawn based on Conditions (i), (ii) and (iii), respectively:

(i) The equation $0 = f(x, \xi)$ obtained by setting $\eta = 0$ in (B.26) implicitly defines a unique C^2 function $x = h(\xi)$.

(ii) For any fixed $\xi \in \mathbb{R}^m$, the equilibrium $x_e = h(\xi)$ of the subsystem $\dot{x} = f(x, \xi)$ is globally asymptotically stable (GAS) and locally exponentially stable (LES).

(iii) The equilibrium $\xi = 0$ of the reduced model (slow timescale)

$$\dot{\xi} = -K'_{g\varepsilon} g(h(\xi), \xi) \tag{B.27}$$

is GAS and LES.

Both conclusions (i) and (ii) are obvious. Only conclusion (iii), the stability (GAS and LES) of the *slow timescale*, needs to be proved. Assume that $K'_{g\varepsilon} > 0$ at first. Consider

$$V(\xi) = \frac{1}{2} \xi^T K'^{-1}_{g\varepsilon} \xi \tag{B.28}$$

as a Lyapunov function candidate for the slow timescale. It can be seen that

$$\dot{V}(\xi) = -\frac{1}{2} \left[\xi^T K'^{-1}_{g\varepsilon} K'_{g\varepsilon} g(h(\xi), \xi) + g^T(h(\xi), \xi) K'_{g\varepsilon} K'^{-1}_{g\varepsilon} \xi \right]$$
$$= -g^T(h(\xi), \xi)\xi = -\xi^T g(h(\xi), \xi). \tag{B.29}$$

This will satisfy the requirements for GAS and LES, given that

$$u_1^T g(h(u_1), u_1) > 0 \tag{B.30}$$

(when $u_1 \neq 0$) and

$$u_1^T g(h(u_1), u_1) \geq \rho |u_1|^2 \tag{B.31}$$

(for some scalar $\rho > 0$) for u_1 in a neighbourhood of $u_1 = 0$. Note that the GAS and LES of the slow timescale are guaranteed for all

$$K'_{g\varepsilon} = \text{diag} \{k'_i \varepsilon_i\} = \text{diag} \left\{ \frac{k_i \varepsilon_i}{\eta} \right\}, \quad 0 < \varepsilon_i \leq 1, \quad i = 1, \cdots, m. \tag{B.32}$$

Now, consider the case that $K'_{g\varepsilon} \geq 0$. In this case, some diagonal elements of $K'_{g\varepsilon}$ are zero and the corresponding controller loops are removed. Without loss of generality, assume that $\varepsilon_j = \varepsilon_l = 0$, where $1 \leq j < l \leq m$. Denote

$$\bar{\xi} = [\xi_1, \cdots, \xi_{j-1}, \xi_{j+1}, \cdots, \xi_{l-1}, \xi_{l+1}, \cdots \xi_m]^T, \tag{B.33}$$

$$\tilde{\xi} = [\xi_1, \cdots, \xi_{j-1}, 0, \xi_{j+1}, \cdots, \xi_{l-1}, 0, \xi_{l+1}, \cdots \xi_m]^T. \tag{B.34}$$

$$\bar{g}(h(\tilde{\xi}), \tilde{\xi}) = \left[g_1\left(h\left(\tilde{\xi}\right), \tilde{\xi}\right), \cdots, g_{j-1}\left(h\left(\tilde{\xi}\right), \tilde{\xi}\right), g_{j+1}\left(h\left(\tilde{\xi}\right), \tilde{\xi}\right), \cdots, \right.$$
$$\left. g_{l-1}\left(h\left(\tilde{\xi}\right), \tilde{\xi}\right) g_{l+1}\left(h\left(\tilde{\xi}\right), \tilde{\xi}\right), \cdots, g_m\left(h\left(\tilde{\xi}\right), \tilde{\xi}\right) \right]^T, \tag{B.35}$$

and

$$\bar{K}'_{g\varepsilon} = \text{diag}\left\{k'_1\varepsilon_1, \cdots, k'_{j-1}\varepsilon_{j-1}, k'_{j+1}\varepsilon_{j+1}, \cdots, k'_{l-1}\varepsilon_{l-1}, k'_{l+1}\varepsilon_{l+1}, \cdots, k'_m\varepsilon_m\right\}. \tag{B.36}$$

Then, the model of slow timescale can be rewritten as

$$\dot{\bar{\xi}} = -\bar{K}'_{g\varepsilon}\bar{g}(h(\tilde{\xi}), \tilde{\xi}). \tag{B.37}$$

Choose

$$\bar{V}(\bar{\xi}) = \frac{1}{2}\bar{\xi}^T \bar{K}'^{-1}_{g\varepsilon}\bar{\xi} \tag{B.38}$$

as a Lyapunov function candidate for the slow timescale model (B.37), then,

$$\dot{\bar{V}}(\bar{\xi}) = -\bar{\xi}^T \bar{g}(h(\tilde{\xi}), \tilde{\xi}). \tag{B.39}$$

The condition $u_1^T g(h(u_1), u_1) > 0$ (when $u_1 \neq 0$) implies that

$$\bar{\xi}^T \bar{g}(h(\tilde{\xi}), \tilde{\xi}) > 0 \tag{B.40}$$

(when $\bar{\xi} \neq 0$), and $u_1^T g(h(u_1), u_1) \geq \rho|u_1|^2$ (for some scalar $\rho > 0$) for u_1 in a neighbourhood of $u_1 = 0$ implies that

$$\bar{\xi}^T \bar{g}(h(\tilde{\xi}), \tilde{\xi}) \geq \rho'|\bar{\xi}|^2 \tag{B.41}$$

(for some scalar $\rho' > 0$) for $\bar{\xi}$ in a neighbourhood of $\bar{\xi} = 0$. Therefore, GAS and LES of the slow timescale model (B.37) can be proved. Then, the conclusion of this theorem follows from Theorem 3.18 in [70].

B.4 Region of Steady-state Attainability

B.4.1 Nominal Stability of Nonlinear IMC

Definition B.5 ([67]). *A continuous function $\upsilon(u) : [0, a) \rightarrow [0, \infty)$ is said to belong to class $\mathcal{K}$ if it is strictly increasing and $\upsilon(0) = 0$.*

Definition B.6 ([67]). *A continuous function $\beta(x, t) : [0, a) \times [0, \infty) \rightarrow [0, \infty)$ is said to belong to class $\mathcal{KL}$ if, for each fixed t, the mapping $\beta(x, t)$ belongs to class $\mathcal{K}$ with respect to x and, for each fixed x, the mapping $\beta(x, t)$ is decreasing with respect to t and $\beta(x, t) \rightarrow 0$ as $t \rightarrow \infty$.*

Definition B.7 ([67]). *Consider the system represented by (6.22) with steady-state equilibrium (u_{ss}, x_{ss}, y_{ss}). The system is said to be input-to-state stable if there exist a class $\mathcal{KL}$ function $\beta(x, t)$ and a class $\mathcal{K}$ function $\upsilon(u)$ such that for any initial state $x_0 \in X_0 \subset \mathbb{R}^n$ and any bounded input $u(t)$ the process state vector $x(t)$ satisfies*

$$\|x(t) - x_{ss}\| \leq \beta(\|x_0 - x_{ss}\|, t - t_0) + \upsilon\left(\sup_{t_0 \leq \tau \leq t} \|u(\tau) - u_{ss}\|\right). \tag{B.42}$$

Based on these definitions, we have the following result:

Theorem B.8 (Stability condition for nonlinear IMC [98]). *Consider the IMC scheme in Figure 6.2, where G, $\tilde{G}$ and Q are nonlinear dynamic systems. Assume that*

1. *the process G is input-to-state stable on $X_0 \subset \mathbb{R}^n$;*
2. *$x_Q \in \mathbb{R}^q$ is the state vector of the IMC controller Q. Moreover, Q is asymptotically stable on $X_Q \subset \mathbb{R}^q$;*
3. *for each steady-state operating point $(e_{ss}, x_{Qss}, u_{ss})$ of Q such that $x_{Qss} \in X_Q$,there exists a corresponding steady-state operating point (u_{ss}, x_{ss}, y_{ss}) of G such that $x_{ss} \in X_0$;*
4. *there is no uncertainty in the process model, i.e., $\tilde{G} = G$;*
5. *the process G and the model $\tilde{G}$ have the same initial conditions $x_0 \in X_0$.*

Then, the IMC closed loop in Figure 6.2 is asymptotically stable on the region $X \triangleq X_0 \times X_Q$ (where $\times$ stands for the Cartesian product).

Proof. From Conditions 4 and 5, $y_0(t) = y(t)$ for all $t > 0$. Thus, the signal $y(t) - y_0(t)$ fed back to the IMC controller in Figure 6.2 is identically zero. Then, $e(t) = r(t)$ for all $t \geq 0$, hence the IMC closed loop in Figure 6.2 is given by

$$y(t) = GQ\, r(t), \quad \forall\, t \geq 0, \tag{B.43}$$

where GQ denotes the series interconnection of the nonlinear dynamic systems G and Q. With $r(t) = r_{ss}$ constant, Condition 2 implies that

$$\begin{aligned} \lim_{t \to \infty} x_Q(t) &= x_{Qss}, \\ \lim_{t \to \infty} u(t) &= u_{ss}, \end{aligned} \tag{B.44}$$

for every initial state in $X_Q \subset \mathbb{R}^q$. From Condition 1, there exist a class $\mathcal{KL}$ function $\beta(x, t)$ and a class $\mathcal{K}$ function $\upsilon(u)$ such that for any initial state $x_0 \in X_0 \subset \mathbb{R}^n$ and any bounded input $u(t)$, the process state vector $x(t)$ satisfies the following condition [67]:

$$\|x(t) - x_{ss}\| \leq \beta(\|x_0 - x_{ss}\|, t - t_0) + \upsilon\left[\sup_{t_0 \leq \tau \leq t} \|u(\tau) - u_{ss}\|\right] \tag{B.45}$$

where the existence of the process steady-state operating point (u_{ss}, x_{ss}, y_{ss}) is guaranteed by Condition 3. The above inequality implies that for any bounded input $u(t)$, the process state vector $x(t)$ will also be bounded. From B.44, we see that the input $u(t)$ converges asymptotically to the steady-state value u_{ss}. This fact, in conjunction with inequality B.42 and the properties of the class $\mathcal{K}$ function $\upsilon(\cdot)$ and class $\mathcal{KL}$ function $\beta(\cdot, \cdot)$, implies that the process state vector $x(t)$ converges asymptotically to x_{ss}. We conclude that the IMC closed loop in B.43 is asymptotically stable in the region $X \triangleq X_0 \times X_Q$.

B.4.2 Proof of Theorem 6.6

Proof ([98]). Based on the process nonlinear model in (6.22) and Assumption 1 in Theorem 6.6, the dynamics of the closed loop in Figure 6.6 are described by the following set of differential equations:

$$\begin{cases} \dot{\xi} &= \varepsilon \hat{K} \left[r_{ss}^\star - g(x, C\,z + D\,\xi) \right] \\ \dot{z} &= A\,z + B\,\xi \\ \dot{x} &= f(x, C\,z + D\,\xi). \end{cases} \tag{B.46}$$

We introduce the *slow* timescale variable τ given by

$$\tau = \varepsilon\,t \tag{B.47}$$

and rewrite the closed-loop dynamics in (B.46) in terms of τ as follows:

$$\begin{cases} \frac{\mathrm{d}\xi}{\mathrm{d}\tau} &= \hat{K} \left[r_{ss}^\star - g(x, C\,z + D\,\xi) \right] \\ \varepsilon\frac{\mathrm{d}z}{\mathrm{d}\tau} &= A\,z + B\,\xi \\ \varepsilon\frac{\mathrm{d}x}{\mathrm{d}\tau} &= f(x, C\,z + D\,\xi). \end{cases} \tag{B.48}$$

The above set of equations are in standard *singular perturbation form* [70]. We then use singular perturbation analysis to study the asymptotic stability of the feasible operating point of interest $(u_{ss}^\star, y_{ss}^\star)$ where $y_{ss}^\star = r_{ss}^\star$. Define the mapping from ξ to y as $y = \varphi(\xi)$. Note that from Assumptions 3 and 4 in Theorem 6.6, the operating point $(u_{ss}^\star, y_{ss}^\star)$ defines a unique operating point $(\xi_{ss}^\star, y_{ss}^\star)$, where $y_{ss}^\star = \varphi(\xi_{ss}^\star)$. Using singular perturbation analysis, we let $\varepsilon \to 0$ and we restrict the analysis to the *slow* timescale dynamics of the closed loop. This is equivalent to a timescale separation approach: By detuning the controller (when $\varepsilon \to 0$), we can separate the "slow" dynamics of the integrator from the "fast" dynamics (in relative terms) of the controller and the process. Thus, when $\varepsilon \to 0$ the closed-loop dynamics in (B.48) become,

$$\begin{cases} \frac{\mathrm{d}\xi}{\mathrm{d}\tau} &= \hat{K} \left[r_{ss}^\star - g(x, C\,z + D\,\xi) \right] \\ 0 &= A\,z + B\,\xi \\ 0 &= f(x, C\,z + D\,\xi), \end{cases} \tag{B.49}$$

for which we require Assumptions 1 and 2. Next, using Assumptions 3 and 4 we can substitute for the above set of equations the following single equation:

$$\frac{\mathrm{d}\xi}{\mathrm{d}\tau} = \hat{K} \left[r_{ss}^\star - \varphi(\xi) \right]. \tag{B.50}$$

To study the stability of the above dynamics, we consider the Lyapunov function candidate

$$V(\xi) = \frac{1}{2} \left(\xi - \xi_{ss}^\star \right)^T \hat{K}^{-1} \left(\xi - \xi_{ss}^\star \right). \tag{B.51}$$

Then,

$$
\frac{\mathrm{d}V}{\mathrm{d}\tau} = \frac{1}{2} \left[\frac{\mathrm{d}\xi}{\mathrm{d}\tau}^T \hat{K}^{-1}(\xi - \xi_{ss}^\star) + (\xi - \xi_{ss}^\star)^T \hat{K}^{-1} \frac{\mathrm{d}\xi}{\mathrm{d}\tau} \right] \tag{B.52}
$$
$$
= - (\xi - \xi_{ss}^\star)^T \left[\varphi(\xi) - r_{ss}^\star \right],
$$

where we have used the fact that $\hat{K}^T = \hat{K}$. We conclude that a sufficient condition that guarantees the asymptotic stability of the equilibrium point $(\xi_{ss}^\star, y_{ss}^\star)$ in the *slow timescale domain* is that $(\xi - \xi_{ss}^\star)^T (y - y_{ss}^\star) > 0$ for every closed-loop trajectory $\xi(\tau)$. The existence of $\varepsilon_0 > 0$ such that the singular perturbation analysis can be applied for all $0 < \varepsilon \leq \varepsilon_0$ is guaranteed by Theorem 3.18 on page 90 of Sepulchre *et al.* [110].

B.4.3 Positive Invariance of Region of Attraction

Proposition B.9 ([98]). *The steady-state region of attraction under linear feedback control $\Omega_u(u_{ss}^\star)$ in Definition 6.7 is positively invariant with respect to $u(t)$ for all $0 < \varepsilon \leq \varepsilon_0$. In addition, every closed-loop trajectory starting from a steady-state initial condition (u_{ss}, y_{ss}), such that $u_{ss} \in \Omega_u(u_{ss}^\star)$, will converge to the feasible operating point $(u_{ss}^\star, y_{ss}^\star)$.*

Proof. The steady-state region of attraction under linear feedback control $\Omega_u(u_{ss}^\star)$ is given by the largest ellipsoid $\Pi\left(\hat{K}, \gamma, u_{ss}^\star\right)$ in the u-space that is completely contained in the AIS and in the region Λ_u. Using the linear relation $u = \bar{K}\xi$, we can express the ellipsoid $\Pi\left(\hat{K}, \gamma, u_{ss}^\star\right)$ in terms of ξ in the ξ-space as follows:

$$
\Pi\left(\hat{K}, \gamma, \xi_{ss}^\star\right) \triangleq \left\{ \xi \in \mathbb{R}^m \mid (\xi - \xi_{ss}^\star)^T \hat{K}^{-1}(\xi - \xi_{ss}^\star) \leq \gamma \right\}. \tag{B.53}
$$

We see that the above ellipsoid is defined based on the Lyapunov function $V(\xi)$ in (B.51). Given $\hat{K} = \hat{K}^T > 0$ and $\gamma > 0$, $V(\xi) \leq \gamma/2$ defines the same ellipsoid given by $\Pi\left(\hat{K}, \gamma, \xi_{ss}^\star\right)$ in (B.53). From Definition 6.7, $\hat{K}$ and γ are such that $\Pi\left(\hat{K}, \gamma, \xi_{ss}^\star\right)$ is the largest ellipsoid completely contained in the region Λ_ξ where Condition (6.30) of Theorem 6.6 holds. Thus, $V(\xi) \leq \gamma/2$ (therefore also $\Omega_u(u_{ss}^\star)$) is positively invariant since $\frac{\mathrm{d}V}{\mathrm{d}\tau} < 0$ for every closed-loop trajectory $\xi(\tau)$ originating in $V(\xi) \leq \gamma/2$. In addition, the result in Theorem 6.6 guarantees that the closed-loop trajectory $\xi(\tau)$ will asymptotically converge to the feasible operating point $(\xi_{ss}^\star, \varphi(\xi_{ss}^\star))$.

B.4.4 Proof of Theorem 6.10

Proof. This proof is an extension of [98]. From the Proof of Theorem 6.6 we have that the *slow timescale dynamics* of the closed loop in Figure 6.6 are given by

$$\frac{\mathrm{d}\xi}{\mathrm{d}\tau} = \hat{K} \left[y'_{ss} - \varphi(\xi) \right], \tag{B.54}$$

where the reference signal is $r(t) = y'_{ss}$ constant. Differentiating the error signal $\tilde{e} = r - y$, we obtain

$$\frac{\mathrm{d}\tilde{e}}{\mathrm{d}\tau} = -\frac{\partial\varphi}{\partial\xi}\frac{\mathrm{d}\xi}{\mathrm{d}\tau}, \tag{B.55}$$

since $y = \varphi(\xi)$. Substituting (B.54),

$$\frac{\mathrm{d}\tilde{e}}{\mathrm{d}\tau} = -\frac{\partial\varphi}{\partial\xi}\hat{K}\left[y'_{ss} - \varphi(\xi)\right] = -\frac{\partial\varphi}{\partial\xi}\hat{K}\tilde{e}. \tag{B.56}$$

Next we show that the equilibrium point $\tilde{e} = 0$ is asymptotically stable. Because $\hat{K} = \hat{K}^T > 0$, we can adopt the following Lyapunov function:

$$V(\tilde{e}) = \frac{1}{2}\tilde{e}^T \hat{K}\tilde{e}. \tag{B.57}$$

Hence,

$$\frac{\mathrm{d}V}{\mathrm{d}\tau} = -\tilde{e}^T \hat{K}^T \frac{\partial\varphi}{\partial\xi}\hat{K}\tilde{e}$$

$$= -\tilde{e}^T \hat{K}^T \frac{\partial h}{\partial u}\bar{K}\hat{K}\tilde{e}. \tag{B.58}$$

Because $\hat{K}$ is nonsingular, we conclude that $\frac{\mathrm{d}V}{\mathrm{d}\tau} < 0$ if and only if

$$\frac{\partial h}{\partial u}\bar{K} > 0. \tag{B.59}$$

If there exists a non-empty region $\Theta_u \subset \mathbb{R}^m$ in the input space such that the above condition is satisfied for every closed-loop trajectory, then the equilibrium $\tilde{e} = 0$ is asymptotically stable.

References

1. A. A. Alonso, C. V. Fernandez, and J. R. Banga. Dissipative systems: From physics to robust nonlinear control. *Int. J. Robust Nonlinear Control*, 14:157–159, 2004.

2. A. A. Alonso and B. E. Ydstie. Process systems, passivity and the second law of thermodynamics. *Comput. Chem. Eng.*, 20(Supplement 2):S1119–S1124, 1996.

3. A. A. Alonso and B. E. Ydstie. Stabilization of distributed systems using irreversible thermodynamics. *Automatica*, 37(11):1739–1755, 2001.

4. A. Amir-Moez. Extreme properties of a Hermitian tranformation and singular values of the sum and product of linear transformation. *Duke Math. J.*, 23:463–476, 1956.

5. B. D. O. Anderson, T. S. Brinsmead, D. Liberzon, and A. S. Morse. Multiple model adaptive control with safe switching. *Int. J. Robust Nonlinear Control*, 8:446–470, 2001.

6. B. D. O. Anderson and S. Vongpanitlerd. *Network Analysis and Synthesis : A Modern Systems Theory Approach*. Prentice-Hall, Englewood Cliffs, N.J., 1973.

7. P. A. Bahri, J. Bandoni, and J. Romagnoli. Integrated flexibility and controllability analysis in design of chemical processes. *AIChE J.*, 42:997–1015, 1997.

8. J. Bao. *Robust Process Control: A Passivity Theory Approach*. PhD thesis, The University of Queensland, 1998.

9. J. Bao, P. L. Lee, F. Y. Wang, and W. B. Zhou. New robust stability criterion and robust controller synthesis. *Int. J. Robust Nonlinear Control*, 8(1):49–59, 1998.

10. J. Bao, P. L. Lee, F. Y. Wang, and W. B. Zhou. Robust process control based on the passivity theorem. *Dev. Chem. Eng. Mineral Process*, 11(3/4):287–308, 2003.

11. J. Bao, P. L. Lee, F. Y. Wang, W. B. Zhou, and Y. Samyudia. A new approach to decentralised process control using passivity and sector stability conditions. *Chem. Eng. Commun.*, 182:213–237, 2000.

12. J. Bao, P. J. McLellan, and J. F. Forbes. A passivity-based analysis for decentralized integral controllability. *Automatica*, 38(2):243–247, 2002.

13. J. Bao, F. Y. Wang, P. L. Lee, and W. B. Zhou. New frequency-domain phase-related properties of MIMO LTI passive systems and robust controller

synthesis. In *Proceedings of IFAC 13th Triennial World Congress*, pages 405–410, San Francisco, 1996.

14. J. Bao, W. Z. Zhang, and P. L. Lee. A passivity-based approach to multi-loop PI controller tuning. In *Proceedings of Sixth International Conference on Control, Automation, Robotics and Vision*, Singapore, 2000. (Paper 178).

15. J. Bao, W. Z. Zhang, and P. L. Lee. A new pairing method for multi-loop control based on the passivity theorem. In *Proceedings of International Symposium on Advanced Control of Industrial Processes*, pages 545–550, Kumamoto, 2002.

16. J. Bao, W. Z. Zhang, and P. L. Lee. Passivity-based decentralized failure-tolerant control. *Ind. Eng. Chem. Res.*, 41(23):5702–5715, 2002.

17. J. Bao, W. Z. Zhang, and P. L. Lee. Decentralized fault-tolerant control system design for unstable processes. *Chem. Eng. Sci.*, 58(22):5045–5054, 2003.

18. G. W. Barton, W. K. Chan, and J. D. Perkins. Interaction between process design and process control: the role of open-loop indicators. *J. Process Control*, 1:161–170, 1991.

19. D. S. Bernstein and W. M. Haddad. LQG control with an H_∞ performance bound: A Riccati equation approach. *IEEE Trans. Automat. Contr.*, 34:293–305, 1989.

20. M. Blanke, M. Kinnaert, J. Lunze, and M. Staroswiecki. *Diagnosis and Fault-tolerant Control*. Springer, Berlin, 2003.

21. S. P. Boyd, L. E. Ghaoui, E. Feron, and V. Balakrishnan. *Linear Matrix Inequalities in System and Control Theory*. SIAM, Philadelphia, 1994.

22. S. P. Boyd and L. Vandenberghe. *Convex Optimization*. Cambridge University Press, 2004.

23. E. H. Bristol. On a new measure of interaction for multivariable process control. *IEEE Trans. Automat. Contr.*, 11(1):133–134, 1966.

24. C. I. Byrnes, A. Isidori, and J. C. Willems. Passivity, feedback equivalence and the global stabilization of minimum phase nonlinear systems. *IEEE Trans. Automat. Contr.*, 36(11):1228–1240, 1991.

25. H. B. Callen. *Thermodynamics and an Introduction to Thermostatistics*. Wiley, New York, 1980.

26. P. J. Campo and M. Morari. Achievable closed-loop properties of systems under decentralized control - conditions involving the steady-state gain. *IEEE Trans. Automat. Contr.*, 39(5):932–943, 1994.

27. Y. Y. Cao, J. Lam, and Y. X. Sun. Static output feedback stabilization: an LMI approach. *Automatica*, 34:1641–1645, 1998.

28. K. H. Chan, J. Bao, and W. J. Whiten. A new approach to control of MIMO processes with static nonlinearities using an extended IMC framework. *Comput. Chem. Eng.*, 30(2):329–342, 2005.

29. X. Chen and J. T. Wen. Positive real controller design with H_∞ norm performance bound. In *Proceedings of American Control Conference*, pages 671–675, Baltimore, 1994.

30. T. P. Chiang and W. L. Luyben. Comparison of energy consumption in five heat-integrated distillation configurations. *Ind. Eng. Chem. Proc. Des. Dev.*, 22(2):175–179, 1983.

31. M. S. Chiu and Y. Arkun. Decentralized control structure selection based on integrity considerations. *Ind. Eng. Chem. Res.*, 29:369–373, 1990.

32. C. A. Desoer and M. Vidyasagar. *Feedback Systems: Input-output Properties*. Academic Press, New York, 1975.

33. C. A. Desoer and Y. T. Wang. Foundations of feedback theory for non-linear dynamical-systems. *IEEE Trans. Circ. Syst.*, 27(2):104–123, 1980.

34. J. M. Douglass. *Conceptual Design of Chemical Processes*. McGraw-Hill, New York, 1988.

35. J. C. Doyle, K. Glover, P. P. Khargonekar, and B. A. Francis. State-space solutions to standard H_2 and H_∞ control problems. *IEEE Trans. Automat. Contr.*, 34:831–846, 1989.

36. S. A. Eker and M. Nikolaou. Linear control of nonlinear systems: Interplay between nonlinearity and feedback. *AIChE J.*, 48(9):1957–1980, 2002.

37. E. Eskinat, S. H. Johnson, and W. L. Luyben. Use of Hammerstein models in identification of nonlinear-systems. *AIChE J.*, 37(2):255–268, 1991.

38. C. A. Farschman, K. P. Viswanath, and B. E. Ydstie. Process systems and inventory control. *AIChE J.*, 44(8):1841–1857, 1998.

39. M. Feinberg. *Lectures on Chemical Reaction Networks*. University of Wisconsin, 1979.

40. J. S. Freudenberg and D. P. Looze. Right half plane poles and zeros and design tradeoffs in feedback systems. *IEEE Trans. Automat. Contr.*, 30:555–565, 1985.

41. P. Gahinet and P. Apkarian. A linear matrix inequality approach to H_∞ control. *Int. J. Robust Nonlinear Control*, 4:421–448, 1994.

42. J. C. Geromel and P. B. Gapski. Synthesis of positive real H_2 controllers. *IEEE Trans. Automat. Contr.*, 42:988–992, 1997.

43. P. Glansdorff and I. Prigogine. *Thermodynamic Theory of Structure, Stability and Fluctuations*. Wiley Interscience, New York, 1971.

44. K. C. Goh and M. G. Safonov. Robustness analysis, sectors and quadratic functions. In *Proceedings of IEEE Conference on Decision and Control*, pages 1988–1993, New Orleans, 1995.

45. A. N. Gorban, I. V. Karlin, and A. Y. Zinovyev. Invariant grids for reaction kinetics. *Physica A*, 33:106–154, 2004.

46. M. Green and D. J. N. Limebeer. *Linear Robust Control*. Prentice-Hall, Upper Saddle River, N.J., 1995.

47. P. Grosdidier and M. Morari. Interaction measures for systems under decentralized control. *Automatica*, 22(3):309–319, 1986.

48. M. Guay, P. J. McLellan, and D. W. Bacon. Measurement of nonlinearity in chemical process control systems: the steady state map. *Can. J. Chem. Eng.*, 73(6):868–882, 1995.

49. E. A. Guillemin. *Synthesis of Passive Networks*. Wiley, New York, 1957.

50. W. M. Haddad and D. S. Bernstein. Robust stabilization with positive real uncertainty: beyond the small gain theory. *Syst. Contr. Lett.*, 17(3):191–208, 1991.

51. W. M. Haddad, D. S. Bernstein, and Y. W. Wang. Dissipative H_2/H_∞ controller synthesis. *IEEE Trans. Automat. Contr.*, 39(4):827–831, 1994.

52. K. M. Hangos, A. A. Alonso, J. D. Perkins, and B. E. Ydstie. A thermodynamical approach to the structural stability of process plants. *AIChE J.*, 45:802–816, 1999.

53. K. M. Hangos, J. Bokor, and G. Szederkényi. Hamiltonian view of process systems. *AIChE J.*, 47:1819–1831, 2001.

54. K. M. Hangos, J. Bokor, and G. Szederkényi. *Analysis and Control of Nonlinear Process Systems*. Springer-Verlag, London, 2004.

55. K. M. Hangos and I. T. Cameron. *Process Modelling and Model Analysis*. Academic Press, London, 2001.

56. N. Hernjak, F. I. Doyle, B. Ogunnaike, and R. Pearson. *Integration of Design and Control*, chapter Chemical Process Characterization for Control Design. Elsevier, 2004.

57. D. Hill and P. Moylan. The stability of nonlinear dissipative systems. *IEEE Trans. Automat. Contr.*, 21:708–711, 1976.

58. M. Hovd and S. Skogestad. Improved independent design of robust decentralized controllers. *J. Process Control*, 3(1):43–51, 1993.

59. M. Hovd and S. Skogestad. Pairing criteria for decentralized control of unstable plants. *Ind. Eng. Chem. Res.*, 33:2134–2139, 1994.

60. S. P. Huang, M. Ohshima, and I. Hashimoto. Dynamic interaction and multiloop control system design. *J. Process Control*, 4(1):15–27, 1994.

61. A. Isidori. *Nonlinear Control Systems*. Springer, London; New York, 3rd edition, 1995.

62. T. Iwasaki and R. E. Skelton. All controllers for the general H_∞ control problem: LMI existence conditions and state space formulas. *Automatica*, 30:1307–1317, 1994.

63. N. Jensen, D. G. Fisher, and S. L. Shah. Interaction analysis in multivariable control systems. *AIChE J.*, 32(6):959–970, 1986.

64. K. Jillson and B. E. Ydstie. Complex process networks: Passivity and optimality. In *Proceedings of 15th IFAC World Congress*, Prague, 2005. (Paper Th-E20-T0/2).

65. D. Kern. *Process Heat Transfer*. McGraw-Hill, 1950.

66. H. K. Khalil. *The Control Handbook*, chapter Stability, pages 889–908. CRC Press, 1995.

67. H. K. Khalil. *Nonlinear Systems*. Prentice-Hall, Upper Saddle River, N.J., 1996.

68. P. Khargonekar and M. Rotea. Mixed H_2/H_∞ control: A convex optimisation approach. *IEEE Trans. Automat. Contr.*, 39:824–837, 1991.

69. P. Kokotović and M. Arcak. Constructive nonlinear control: a historical perspective. *Automatica*, 37:637–662, 2001.

70. P. Kokotović, H. Khalil, and J. O'Reilly. *Singular Perturbation Methods in Control: Analysis and Design*. Academic Press, Orlando, 1986.

71. I. K. Kookos and J. D. Perkins. An algorithm for simultaneous process design and control. *Ind. Eng. Chem. Res.*, 40(19):4079–4088, 2001.

72. H. J. Kreutzer. *Nonequilibrium Thermodynamics and Its Statistical Foundations*. Clarendon Press, New York, 1983.

73. H. Kwakernaak and R. Sivan. *Linear Optimal Control Systems*. Wiley Intersciences, New York, 1972.

74. S. Lakshminarayanan, S. I. Shah, and K. Nandakumar. Identification of Hammerstein models using multivariate statistical tools. *Chem. Eng. Sci.*, 50(22):3599–3613, 1995.

75. D. K. Le, O. D. I. Nwokah, and A. E. Frazho. Multivariable decentralized integral controllability. *Int. J. Control*, 54:481–496, 1991.

76. J. Lee and T. F. Edgar. Conditions for decentralized integral controllability. *J. Process Control*, 12:797–805, 2002.

77. T. K. Lee, J. X. Shen, and M. S. Chiu. Independent design of robust partially decentralized controllers. *J. Process Control*, 11:419–428, 2001.

78. W. Luyben, B. D. Tyréus, and M. L. Luyben. *Plantwide Process Control*. McGraw-Hill, New York, 1999.

79. W. L. Luyben. Simple method for tuning SISO controllers in multivariable systems. *Ind. Eng. Chem. Process Des. Dev.*, 25:654–660, 1986.

80. W. L. Luyben. *Integration of Design and Control*, chapter The need for simultaneous design education. Elsevier, 2004.

81. V. Manousiouthakis, R. Savage, and Y. Arkun. Synthesis of decentralized process-control structures using the concept of block relative gain. *AIChE J.*, 32(6):991–1003, 1986.

82. B. Maschke, A. van der Schaft, and P. Breedveld. An intrinsic Hamiltonian formulation of the dynamics of LC-circuits. *IEEE Trans. Circ. Syst.*, 42:73–82, 1995.

83. P. S. Maybeck and R. D. Stevens. Reconfigurable flight control via multiple model adaptive control methods. *IEEE Trans. Aerosp.Electron. Syst.*, 27:470–480, 1991.

84. T. J. McAvoy. Some results on dynamic interaction analysis of complex control systems. *Ind. Eng. Chem. Process Des. Dev.*, 22:42–49, 1983.

85. D. C. McFarlane and K. Glover. *Robust Controller Design Using Normalized Coprime Factor Plant Descriptions*. Springer, New York, 1990.

86. M. Morari. Design of resilient processing plants - III. a general framework for the assessment of dynamic resilience. *Chem. Eng. Sci.*, 38:1881–1891, 1983.

87. M. Morari and E. Zafiriou. *Robust Process Control*. Prentice-Hall, Englewood Cliffs, N.J., 1989.

88. A. A. Niederlinski. Heuristic approach to the design of linear multivariable interacting control systems. *Automatica*, 7:691–701, 1971.

89. M. Nikolaou and P. Misra. Linear control of nonlinear processes: recent developments and future directions. *Comput. Chem. Eng.*, 27:1043–1059, 2003.

90. R. Ortega, A. van der Schaft, and B. Maschke. Stabilization of port-controlled Hamiltonian systems via energy balancing. *Lecture Notes in Control and Information Sciences*, 246:239–260, 1999.

91. A. Ostrowski and H. Schneider. Some theorems on the inertia of general matrices. *J. Math. Anal. Appl.*, 4:72–84, 1962.

92. I. Otero-Muras, G. Szederkényi, A. Alonso, and K. Hangos. Dynamic analysis and control of chemical and biochemical reaction networks. In *Proceedings of the International Symposium on Advanced Control of Chemical Processes – ADCHEM 2006*, Gramado, 2004.

93. J. D. Perkins. The integration of design and control the key to future processing systems? In *Proceedings of 6th World Congress of Chemical Engineering*, Melbourne, 2001. (Keynote).

94. J. D. Perkins and S. P. K. Walsh. Optimization as a tool for design/control integration. *Comput. Chem. Eng.*, 20(4):315–323, 1996.

95. V. M. Popov. *Hyperstability of Control Systems*. Springer-Verlag, New York, 1973.

96. I. Postlethwaite, J. Edmunds, and A. G. J. MacFarlane. Principal gains and principal phases in the analysis of linear multivariable feedback systems. *IEEE Trans. Automat. Contr.*, 26:32–46, 1981.

97. O. Rojas, J. Bao, and P. Lee. A dynamic operability analysis approach for nonlinear processes. *J. Process Control*, 17(2):157–172, 2007.

98. O. J. Rojas, J. Bao, and P. L. Lee. Linear control of nonlinear processes: the regions of steady-state attainability. *Ind. Eng. Chem. Res.*, 45(22):7552–7565, 2006.

99. H. H. Rosenbrock. *Computer-aided Control System Design*. Academic Press, Orlando, F.L., 1974.

100. P. Rouchon and Y. Creff. Geometry of the flash dynamics. *Chem. Eng. Sci.*, 18:3141–3147, 1993.

101. M. G. Safonov. Stability margins for diagonally perturbed multivariable feedback systems. *IEE Proc. Part D*, 129:251–256, 1982.

102. M. G. Safonov, E. A. Jonckheere, M. Verma, and D. J. N. Limebeers. Synthesis of positive real multivariable feedback systems. *Int. J. Control*, 45:817–842, 1987.

103. V. Sakizlis, J. D. Perkins, and E. N. Pistikopoulos. Recent advances in optimization-based simultaneous process and control design. *Comput. Chem. Eng.*, 28(10):2069–2086, 2004.

104. Y. Samyudia, P. L. Lee, I. T. Cameron, and M. Green. Control strategies for a supercritical fluid extraction process. *Chem. Eng. Sci.*, 51(5):769–787, 1996.

105. H. Santoso, J. Bao, P. L. Lee, and O. J. Rojas. Dynamic controllability analysis for multi-unit processes. In *Proceedings of CHEMECA 2006*, Auckland, 2006. (Paper 252).

106. H. Santoso, O. J. Rojas, J. Bao, and P. L. Lee. The regions of steady-state attainability for nonlinear processes: A case study. In *Proceedings of CHEMECA 2006*, Auckland, 2006. (Paper 215).

107. C. Scherer, P. Gahinet, and M. Chilali. Multiobjective output-feedback control via LMI optimization. *IEEE Trans. Automat. Contr.*, 42:896–911, 1997.

108. C. Schweiger and C. Floudas. *Optimal Control: Theory, Algorithms and Applications*, chapter Integration of Design and Control: Optimisation with Dynamic Models. Kluwer Academic, Norwell, MA, 1997.

109. W. D. Seider, J. D. Seader, and D. R. Lewin. *Product and Process Design Principles: Synthesis, Analysis, and Evaluation*. Wiley, New York, 2003.

110. R. Sepulchre, M. Janković, and P. Kokotović. *Constructive Nonlinear Control*. Springer, London, New York, 1997.

111. D. D. Siljak. Reliable control using multiple control systems. *Int. J. Control*, 31(2):303–329, 1980.

112. S. Skogestad and M. Morari. Robust performance of decentralized control systems by independent design. *Automatica*, 25(1):119–125, 1989.

113. S. Skogestad and M. Morari. Variable selection for decentralized control. *Modeling Identification and Control*, 13(2):113–125, 1992.

114. S. Skogestad and I. Postlethwaite. *Multivariable Feedback Control - Analysis and Design*. Wiley, Chichester, 2nd edition, 2005.

115. J. J. E. Slotine. Putting physics in control – the example of robotics. *IEEE Control Syst. Mag.*, 8:12–18, 1988.

116. J. J. E. Slotine and W. Li. *Applied Nonlinear Control*. Prentice-Hall, Englewood Cliffs, N.J., 1991.

117. E. Sontag. Structure and stability of certain chemical networks and applications to the kinetic proofreading model of t-cell receptor signal transduction. *IEEE Trans. Automat. Contr.*, 46:1028–1047, 2001.

118. A. J. Stack and F. J. Doyle. The optimal control structure: an approach to measuring control-law nonlinearity. *Comput. Chem. Eng.*, 21(9):1009–1019, 1997.

119. S. W. Su, J. Bao, and P. L. Lee. Analysis of decentralized integral controllability for nonlinear systems. *Comput. Chem. Eng.*, 28(9):1781–1787, 2004.

120. S. W. Su, J. Bao, and P. L. Lee. Passivity based imc control for multivariable nonlinear systems. In *Proceedings of 5th Asian Control Conference*, pages 135–140, Melbourne, 2004.

121. S. W. Su, J. Bao, and P. L. Lee. Conditions on input disturbance suppression for multivariable nonlinear systems on the basis of feed forward passivity. *Int. J. Syst. Sci.*, 37(4):225–233, 2006.

122. S. W. Su, J. Bao, and P. L. Lee. A hybrid active-passive fault tolerant control approach. *Asia-Pac. J. Chem. Eng.*, 1(1/2):54–62, 2007.

123. W. Q. Sun, P. P. Khargonekar, and D. S. Shim. Solution to the positive real control problem for linear time-invariant systems. *IEEE Trans. Automat. Contr.*, 39(10):2034–2046, 1994.

124. A. D. Suryodipuro, J. Bao, and P. L. Lee. Process controllability analysis based on passivity condition. In *Proceedings of CHEMECA 2004*, Sydney, 2004. (Paper 81).

125. A. D. Suryodipuro, J. Bao, and P. L. Lee. Controller structure selection based on achievable dynamic performance. In *Proceedings of CHEMECA 2005*, Brisbane, 2005. (Paper 97).

126. M. Takagaki and S. Arimoto. A new feedback method for dynamic control of manipulators. *ASME J. Dynamic Syst. Meas. Control*, 103:119–125, 1981.

127. K. Tan and K. M. Grigoriadis. Stabilization and H_∞ control of symmetric systems: an explicit solution. *Syst. Contr. Lett.*, 44:57–72, 2001.

128. B. D. Tyreus and W. L. Luyben. Controlling heat integrated distillation columns. *Chem. Eng. Progr.*, 72(9):59–66, 1976.

129. D. Uztürk and C. Georgakis. Inherent dynamic operability of processes: General definitions and analysis of SISO cases. *Ind. Eng. Chem. Res.*, 41:421–432, 2002.

130. A. J. van der Schaft. *L_2-gain and Passivity Techniques in Nonlinear Control*. Springer, London, New York, 2000.

131. M. Vidayasagar and N. Viswanadham. Reliable stabilization using a multi-controller configuration. *Automatica*, 21(5):599–602, 1985.

132. D. R. Vinson and C. Georgakis. A new measure of process output controllability. *J. Process Control*, 10:185–194, 2000.

133. K. Wei. Stabilization of a linear plant via a stable compensator having no real unstable zeros. *Syst. Contr. Lett.*, 15:259–264, 1990.

134. O. Weitz and D. Lewin. Dynamic controllability and resiliency diagnosis using steady state process flowsheet data. *Comput. Chem. Eng.*, 20(4):325–335, 1996.

135. J. T. Wen. Robustness analysis based on passivity. In *Proceedings of American Control Conference*, pages 1207–1212, Atlanta, 1988.

136. J. T. Wen and D. Bayard. New class of control laws for robotic manipulators: Part 1, Non-adaptive case. *Int. J. Control*, 47(5):1361–1385, 1988.

137. V. White, J. Perkins, and D. Espie. Switchability analysis. *Comput. Chem. Eng.*, 20(4):469–474, 1996.

138. J. C. Willems. Dissipative dynamical systems, Part I: General theory. *Arch. Rational Mech. Anal.*, 45(5):321–351, 1972.

139. J. C. Willems. Dissipative dynamical systems, Part II: Linear-systems with quadratic supply rates. *Arch. Rational Mech. Anal.*, 45(5):352–393, 1972.

140. A. S. Willsky. A survey of design methods for failure detection in dynamic systems. *Automatica*, 12:601–611, 1976.

141. D. A. Wilson. Convolution and Hankel operator norms for linear systems. *IEEE Trans. Automat. Contr.*, 34:94–97, 1989.

142. Q. Xia and M. Rao. Fault tolerant control of paper machine headboxes. *J. Process Control*, 2:171–178, 1992.

143. B. E. Ydstie. Passivity based control via the second law. *Comput. Chem. Eng.*, 26:1037–1048, 2002.

144. B. E. Ydstie and A. A. Alonso. Process systems and passivity via the Clausius-Planck inequality. *Syst. Contr. Lett.*, 30(5):253–264, 1997.

145. C. C. Yu and M. K. H. Fan. Decentralized integral controllability and D-stability. *Chem. Eng. Sci.*, 45(11):3299–3309, 1990.

146. C. C. Yu and W. L. Luyben. Design of multiloop SISO controllers in multivariable processes. *Ind. Eng. Chem. Process Des. Dev.*, 25:498–503, 1986.

147. G. Zames. On the input and output stability of time-varying nonlinear feedback systems. *IEEE Trans. Automat. Contr.*, 11:228–476, 1966.

148. W. Z. Zhang. *Analysis and Design of Decentralized Fault-tolerant Control Systems Based on the Passivity Theorem*. PhD thesis, The University of New South Wales, 2003.

149. W. Z. Zhang, J. Bao, and P. L. Lee. Decentralized unconditional stability conditions based on the passivity theorem for multi-loop control systems. *Ind. Eng. Chem. Res.*, 41(6):1569–1578, 2002.

150. W. Z. Zhang, J. Bao, and P. L. Lee. Synthesis of strictly positive real/H_∞ controllers using a linear matrix inequality approach. In *Proceedings of International Symposium on Advanced Control of Industrial Processes*, pages 275–279, Kumamoto, 2002.

151. W. Z. Zhang, J. Bao, and P. L. Lee. Control structure selection based on block-decentralized integral controllability. *Ind. Eng. Chem. Res.*, 42(21):5152–5156, 2003.

152. W. Z. Zhang, J. Bao, and P. L. Lee. Pairing studies of multivariable processes under block decentralized control. In *Proceedings of CHEMECA 2003*, Adelaide, 2003. (Paper 241).

153. W. Z. Zhang, J. Bao, and P. L. Lee. Process dynamic controllability analysis based on all-pass factorization. *Ind. Eng. Chem. Res.*, 44 (18):7175–7188, 2005.

154. D. H. Zhou and P. M. Frank. Fault diagnostics and fault tolerant control. *IEEE Trans. Automat. Contr.*, 34(2):420–427, 1998.

155. K. Zhou and J. C. Doyle. *Essentials of Robust Control*. Prentice-Hall, Upper Saddle River, N.J., 1998.

156. K. Zhou, J. C. Doyle, and K. Glover. *Robust and Optimal Control*. Prentice-Hall, Upper Saddle River, N.J., 1996.

Index

balance equations
    input convection term, 201
    output convection term, 201
    overall mass, 201
    source term, 201
    transfer term, 201
balance volumes, 200
basic modelling assumptions, 207
BDIC
    see block decentralized integral
        controllability, 104
block decentralized integral controlla-
        bility, 103, 104
    necessary condition, 107
    sufficient condition, 105
blocking zeros, 150
bounded-real lemma, 64

Cayley transformation, 61, 74
conserved extensive variables, 200
constitutive equations
    extensive-intensive relationship, 202
    physicochemical property relations,
        203
    reaction rate expression, 202
    source equations, 202
    thermodynamic state equations, 203
    transfer rate equations, 202
controllability, 15
    dynamic controllability, 161
    input-output controllability, 161
controllability analysis
    nonlinear processes, 181, 187
    stable linear systems, 166

steady-state attainability, 171, 179,
        181, 184, 185
steady-state attainability via linear
        feedback control, 173
steady-state output space achievable,
        178, 180
steady-state region of attraction, 172,
        175, 176
costate variables, 212

decentralized control, 89
    achievable performance, 116
    control design based on passivity, 120
    stability condition, 115
decentralized detunability
    see decentralized unconditional
        stability, 126
decentralized integral controllability, 91
    linear systems, 92
    necessary condition, 92
    passivity-based condition, 94, 95
    passivity-based condition for
        nonlinear systems, 98–100
    sufficent condition, 92
decentralized unconditional stability,
        126, 128
    pairing schemes, 132
    passivity-based condition, 128
DIC
    see decentralized integral controllabil-
        ity, 91
dissipation matrix, 214
dissipative systems, 9
    available storage, 10

examples, 7
storage function, 9
supply rate, 9, 27
dynamic interaction measure, 110, 111
    PB-IM, 112, 114
    SB-IM, 115
    SSV-IM, 112

engineering intensive variables, 201
entropic intensive variables, 196
entropy-based storage function, 208
    choice of inputs and outputs, 210
    generalized, 208
    properties, 209
Euler equation, 197
extensive variables, 194
    canonical set, 194
    conserved, 200
    entropy, 194
    relation with the intensive ones, 199
extensive-intensive relationship, 199,
    201

fault-tolerant control, 125
    $\mathcal{H}_2$ control, 139–141, 143, 145
    $\mathcal{H}_\infty$ control, 146
    failure modes and effects analysis, 156
    fault accommodation, 159
    fault detection, 157, 158
    hybrid approach, 156, 159
    PI control, 135
    unstable processes, 149, 150, 152, 153
    virtual sensor, 159
first law of thermodynamics, 197
FTC
    see fault-tolerant control, 125

Hamiltonian function, 213
Hamiltonian storage function, 213
Hamiltonian systems, 212
    generalized, 213
    states and costates, 212
    with dissipation, 214

intensive variables, 195
    driving forces, 196
    engineering, 201
    entropic, 196
    mass specific extensive, 201

pressure, 195
    relation with the extensive ones, 199
    temperature, 195
internal model control, 164
    extended IMC, 166

KYP lemma
    see positive-real lemma, 14
KYP property, 13

laws of thermodynamics, 197
    first, 197
    second, 198
Lie derivative, 13
local thermodynamic equilibrium, 198
lossless systems, 213

mechanisms, 201
    convection, 201
    transfer, 201
minimum phase systems, 20

nonequilibrium thermodynamics, 198
    Onsager relation, 198

Onsager relation, 198
    engineering version, 202

passivation
    input feedforward passivation, 29
    output feedback passivation, 30
passive systems, 10
    examples, 7, 36
    interconnection, 21
    lossless systems, 11
    output feedback stability, 21
    partial interconnection, 22
    phase condition, 18, 20
    stability, 12
    state strictly passive systems, 11
passivity
    excess, 25
    incremental input passivity, 179
    incrementally strict input passivity,
        179
    shortage, 25
passivity index, 24
    diagonal scaling, 113, 114, 130, 131
    frequency-dependent IFP, 28
    frequency-dependent OFP, 29

IFP, 24–26
IFP of linear systems, 28
OFP, 25, 26
OFP of linear systems, 29
passivity theorem, 32, 35
   linear version, 34
passivity-based stability condition, 32, 34, 35
positive real systems, 15
   linear extended strictly positive real systems, 17
   linear positive real systems, 17
   linear strictly positive real systems, 17
   positive real over a frequency band, 70
positive-real lemma, 14
process modelling
   balance volumes, 200
   conservation balances, 201
   mechanisms, 201
   overall mass balance, 201

reachability, 15
relative degree, 19
robust control, 56
   nominal performance, 57, 75
   passive controller design, 80, 85
   passivity-based control design, 59, 64, 73, 74, 77
   passivity-based stability condition, 56
robust stability, 46
   passivity and gain condition, 70, 72
   passivity-based condition, 50, 56
   small gain condition, 46

second law of thermodynamics, 198
sector, 52
   properties, 53
   sector stability condition, 53
signal norm, 15
space
   $\mathcal{L}_{2e}$ space, 16

$\mathcal{L}_2$ space, 16
stability
   asymptotic stability, 6
   exponential stability, 7
   global stability, 6
   input-output stability, 16
   locally exponential stability, 7
   Lyapunov stability, 6
   stability criterion, 6
state equations, 203
   intensive form, 203, 210
   structure, 204
static feedback stabilizability, 150
storage function, 207
   entropy-based, 208
   entropy-based, generalized, 208

thermodynamic driving forces, 196
thermodynamics, 194
   entropic intensive variables, 196
   equations of state, 196
   equilibrium, 198
   extensive variables, 194
   fundamental equation, 195
   intensive variables, 195
   nonequilibrium, 198
   postulates, 194

uncertainty, 44
   additive uncertainty, 44
   IFP bound, 48, 49
   multiplicative uncertainty, 45
   parameter uncertainty, 44
   passivity-based uncertainty measure, 48, 49, 70
   sector bounded passivity measure, 53–55
   simultaneous IFP OFP bounds, 51
   unstructured uncertainty, 44

zero dynamics, 19
zero-state detectability, 12
zero-state observability, 12

Other titles published in this Series (continued):

Adaptive Voltage Control in Power Systems
Giuseppe Fusco and Mario Russo

Advanced Control of Industrial Processes
Piotr Tatjewski

Process Control Performance Assessment
Andrzej Ordys, Damien Uduehi and
Michael A. Johnson (Eds.)

*Modelling and Analysis of Hybrid
Supervisory Systems*
Emilia Villani, Paulo E. Miyagi and
Robert Valette

*Continuous-time Model Identification
from Sampled Data*
Hugues Garnier and Liuping Wang (Eds.)
Publication due September 2007

Distributed Embedded Control Systems
Matjaž Colnarič, Domen Verber and
Wolfgang A. Halang
Publication due October 2007

Model-based Process Supervision
Belkacem Ould Bouamama and
Arun K. Samantaray
Publication due November 2007

Optimal Control of Wind Energy Systems
Iulian Munteanu, Antoneta Iuliana
Bratcu, Nicolas-Antonio Cutululis and
Emil Ceanga
Publication due November 2007

*Model Predictive Control Design and
Implementation Using MATLAB®*
Liuping Wang
Publication due November 2007

*Dry Clutch Control for Automated Manual
Transmission Vehicles*
Pietro J. Dolcini, Carlos Canudas-de-Wit
and Hubert Béchart
Publication due February 2008

Magnetic Control of Tokamak Plasmas
Marco Ariola and Alfredo Pironti
Publication due February 2008